John Call Dalton

The experimental method in medical science.

Second course of the Cartwright lictures of the Alumni association, College of physicians and surgeons, New York, delivered January 24, January 31, and February 7, 1882

John Call Dalton

The experimental method in medical science.
Second course of the Cartwright lictures of the Alumni association, College of physicians and surgeons, New York, delivered January 24, January 31, and February 7, 1882

ISBN/EAN: 9783337873127

Printed in Europe, USA, Canada, Australia, Japan

Cover: Foto ©berggeist007 / pixelio.de

More available books at **www.hansebooks.com**

THE

EXPERIMENTAL METHOD

IN

MEDICAL SCIENCE

SECOND COURSE OF THE CARTWRIGHT LECTURES OF THE
ALUMNI ASSOCIATION, COLLEGE OF PHYSICIANS
AND SURGEONS, NEW YORK

DELIVERED

JANUARY 24, JANUARY 31, AND *FEBRUARY 7, 1882*

BY

JOHN C. DALTON, M.D.

NEW YORK
G. P. PUTNAM'S SONS
27 AND 29 WEST 23D STREET
1882

COPYRIGHT, BY
G. P. PUTNAM'S SONS,
1882.

CONTENTS.

LECTURE I.

GALVANI AND GALVANISM IN THE STUDY OF THE NERVOUS SYSTEM 1

PAGE

LECTURE II.

BUFFON AND BONNET IN THE EIGHTEENTH CENTURY . . . 39

LECTURE III.

NERVOUS DEGENERATIONS AND THE THEORY OF SIR CHARLES BELL 72

THE EXPERIMENTAL METHOD IN MEDICAL SCIENCE.

LECTURE I.

GALVANI AND GALVANISM IN THE STUDY OF THE NERVOUS SYSTEM.

Mr. President and Gentlemen of the Alumni, — In discharging the agreeable duty which you have kindly imposed upon me, of giving the Cartwright Lectures for this year, I propose to offer a few historical sketches, which shall illustrate the manner in which certain parts of our scientific knowledge in medicine have been attained. The connection between scientific and practical medicine is furthermore so close, that permanent improvement in the one is inseparably dependent on that of the other; but this connection becomes much more apparent when we trace the history of any particular department for a considerable period of time. By this means we can see how much of the doctrine accepted by our predecessors has survived the ordeal of a century, and what were the methods of investigation which produced in their hands the permanent results which we enjoy to-day. However much we may pride ourselves on the advances made during our own time, we may be sure that by far the greater part of our actual knowledge is a legacy from the past. It has been winnowed in successive generations from the errors and imperfections which always accompany its first acquisition; and it is probable that many of our own discoveries will require a similar depleting treatment in the future. But we may learn to rely with confidence on such methods of study as have heretofore proved valuable; and we may, perhaps, save ourselves the trouble of exploring certain paths, when we see that others have followed them before, and have found that they lead nowhere.

If we can say that any one department in physiology, pathology, and therapeutics, is now distinguished by a special activity of investigation and growth, it is probably that of the nervous system. A very large part of this advance has been made by the application of electric stimulus in determining the motor or sensitive properties of different nerves, their influence on the heart and blood-vessels, the localization of special centres in the brain and spinal cord, or the diagnosis of morbid alterations in the cerebro-spinal axis; and the use of electricity in restoring the power of movement or sensation when impaired by disease is now acknowledged to be, in many cases, a most serviceable means of cure. This has all come, directly or indirectly, from the experiments of Galvani, nearly a hundred years ago, on the nerves and muscles of the decapitated frog.

In 1789 Galvani was professor of anatomy in the University of Bologna. In addition to the regular duties of his professorial chair, he had made a number of valuable investigations in comparative anatomy, such as those on the structure of the kidneys and urinary ducts of birds, and on the organ of hearing in the same class. Like most of the scientific men of his day, he was also greatly interested in the phenomena of electricity, which was then fast developing into an important department of physics. The condition of electrical science at that time was as follows. The machine for producing frictional electricity by a rotating glass cylinder and cushion had been brought to practical completion, and was in common use, with its prime conductor and insulating supports. The two opposite kinds of electrical excitement, known as vitreous and resinous, or positive and negative, were fully recognized, as well as the distinction between conductors and non-conductors; and even some of the phenomena of induced electricity were known, though explained in a manner somewhat different from that which is now in vogue. Besides the electrical machine, experimenters were already in possession of the Leyden jar, by which a large quantity of electricity may be stored in a given space; the electrophorus, by which a moderate charge of electricity may be obtained at will from a permanent source; several varieties of electrometers, or electroscopes, for detecting the existence and amount of slight electric disturbances; and Volta's "condenser," in which small quantities of electricity from a feeble source might be accumulated, and made apparent by the electrometer. Finally, Franklin had shown, by his daring experiment with the kite in the thunder-storm, that the lightning of the clouds was identical in its nature with the spark from an electrical machine; and this had largely directed the attention of investigators to the study of atmospheric electricity, as compared with that produced by artificial means.

Considering the long list of results which have followed from Galvani's early observations, the manner in which they were first made is a topic of much interest.[1] He was in his laboratory, engaged on experiments with the electrical machine, and had, lying upon the table near by, a freshly dissected frog, prepared for some other purpose in such a way that the denuded hind legs were connected with the spinal column by the crural nerves. One of his assistants, accidentally touching the nerves of the animal with the blade of a scalpel, saw the legs convulsed; and, on watching more closely, it was seen that the contraction occurred only at the moment of drawing a spark from the conductor of the machine. Once Galvani's attention attracted to so remarkable a phenomenon, his mind turned instantly to the investigation of its conditions. He abandoned all other occupations, and seemed absorbed in the attempt to detect its causes, and to learn their mode of operation. He determined, in the first place, that the discharge of the conductor, and the contact of the scalpel with the frog's nerves, were both necessary; for the muscular contraction would not take place with either of them alone. But, even when both conditions were present, sometimes the muscles contracted, and sometimes they did not. Puzzled by this variation, but still confident that it must have a reasonable cause, he at last found that it depended on the way the scalpel was held in the fingers. If grasped by the end of its non-conducting ivory handle, there were no convulsions in the legs when the spark was taken from the machine; but if held in such a way that the fingers touched the steel blade, or the rivets which held it in place, the muscles were always thrown into action. The human body, therefore, served as a con-

[1] Aloysii Galvani de Viribus Electricitatis in Motu Musculari Commentarius. De Bononiensi Scientiarum et Artium Instituto atque Academia Commentarii, 1791. Tomus septimus, p. 363.

ductor; and Galvani replaced it with success by an iron wire,-which he attached by one extremity to the spinal column, above the origin of the nerves, by a brass or copper hook, leaving the other end in communcation with the ground. Then he could dispense with the scalpel altogether. He varied the contrivance in many ways, turning sometimes the attached and sometimes the free end of the conductor toward the electrical machine, increasing or diminishing its length, and at last adding a second conductor attached to the muscles of the leg. He took especial pains to exclude the possibility of any direct transfer of electricity from his machine to the dissected frog, and yet convinced himself, by the aid of Volta's electrometer, that in some way or other an electric discharge passed through the frog, and was the exciting cause of its convulsions.

These experiments were all performed with artificial electricity obtained from the electrical machine. Galvani then passed to his second series of observations, to see whether a similar effect would be produced by atmospheric electricity. On the approach of a thunder-storm, he arranged one of the conducting wires so that its upper extremity was in the open air, near the roof of his house, and its lower extremity connected with the frog's spinal column, while the other wire, attached to the muscles of the leg, communicated below with the water in a well; and he then waited, like Franklin, for the discharge of a thunder-cloud. The result followed as before, and every peal was accompanied by a convulsive motion in the dissected frog. He especially noticed that the convulsions were simultaneous, not with the sound of the thunder, but with the electric discharge; for, as he says, "the muscular contractions and movements of the animal, like the visible splendor of the lightning-flash, always preceded the thunder-clap, and, as it were, gave notice that it was coming."

So far Galvani had not really touched upon his final discovery. The phenomena of muscular contraction in these two sets of experiments were due, as we now know, to the action of induced electricity. When he charged the prime conductor of his electrical machine, or when the thunder-cloud was passing over his house, an imperceptible but real disturbance of the electric equilibrium took place by induction in the wires attached to his dissected frog; and, when the original electrical tension expended itself in a discharge, there was a similar re-action in his conducting wires, — a re-action which necessarily passed through the nerves and muscles of the frog, and threw it into convulsions. Galvani did not understand this; but he was engrossed with the idea of pursuing the new fact which he had unexpectedly witnessed, and of following its investigation by whatever path might be open to him. So it happened that he was led, in his third set of experiments, to a result of far greater importance; and notwithstanding that the two former series had nothing to do with this result, so far as the nature of their phenomena were concerned, they were still the necessary and perfectly natural preliminaries to its discovery.

Galvani knew that the violent disturbances of thunder and lightning were not the only ones which take place in the atmosphere, but that there were also other changes of electrical condition going on more quietly in ordinary weather. "After having tested," he says, "the action of atmospheric electricity in thunder-storms, I was exceedingly desirous of investigating it in its daily condition of quietude and serenity." For this purpose he hung his freshly dissected frogs, by copper hooks attached to the spinal column, from the iron railing of a balustrade about his house, and left them there exposed to the sky. He thought that perhaps the imperceptible exchange of electricity between the clouds and the atmosphere might be betrayed by some movement in the frogs' limbs, and he watched patiently with that object for several days; but there was not the least indication of muscular excitement. At last, seeing that nothing further would come of the experiment in that way, he shifted the position of the limbs, and pressed them, still hanging by their hooks, against the iron framework of the balustrade; and instantly a convulsion took place. He had made a galvanic circuit. The copper hook and the iron railing were the two metals of the battery; and, by bringing the muscles in contact with the railing, he had completed the communication, and had seen, for the first time, a muscular contraction excited by the galvanic current.

But of course his knowledge of the real cause of the contraction was still very incomplete; and no one could be more conscious of this than himself. He knew how many sources of error there may be in drawing a conclusion from visible facts, until they have been examined from every direction. His most natural inference, if he had remained satisfied with this experiment, would have been that the electrical cause of the contractions came from the atmosphere. "I was very much inclined," he says, "to attribute these contractions to atmospheric electricity which had accumulated in the frog, and was then suddenly discharged by the contact of the hooks and railings,—never having seen contractions produced in this way, except in the open air, *as I had not yet tried them in any other place,*—we may be so easily misled into thinking that we have really seen whatever we hope or expect to see." But he soon found that he could obtain the same results in his closed laboratory as in the open air, at any time of the day, in any weather, and under a great variety of conditions; provided, only, that communication was made between the nerves and muscles by metallic conductors, with no intervening non-conducting material. The best arrangement was that in which the nerves were suspended by copper hooks, or armed with tin-foil, while the denuded feet of the frog rested on a silver plate, and the two were then connected by a metallic arc. The human body, being also a conductor, might be interposed between the ends of the arc without lessening the result. Galvani's description of this part of the discovery is an interesting episode, as it corroborates his proof of the electric nature of the force in operation. He found that if he held the frog suspended by its copper hook in one hand, and touched the silver plate with a metallic rod in the other, the frog was convulsed. Then he called in the assistance of a colleague. "I was staying," he says, "at the country-house of the most noble and excellent Signor James Zambeccari, where Signor Rialpo of Spain, formerly a member of the Society of Jesus, and a very learned man, was visiting at the same time. He had already kindly aided me in former experiments, and I begged him to do so again in this instance. He accordingly took my place in holding the frog, while, partly for convenience, and partly to vary the conditions of the experiment, I myself touched the silver plate with my metallic rod; but, contrary to expectation, no movement was produced." Then the two observers joined hands, and, on repeating the metallic contact, found "to their delight" that communication took place through the human electric chain, and was manifested by muscular contraction in the frog's leg. From all these results the conclusion became irresistible, that the convulsions previously observed in the frogs attached to the iron railings were due to the contact of their metallic supports.

It is perfectly evident that this third set of experiments contained the substance of a new discovery, and Galvani fully recognized the fact. In the two former series there were external sources of electricity, either in the electrical machine or in the thunder-clouds of the atmosphere. But here all such conditions were excluded. There was nothing to explain the phenomena but the frog's nerves and muscles, and the metallic conductors between them. In all Galvani's experiments his mind was forcibly impressed by this fact, and he was naturally led to suspect that the electricity which caused the convulsions might be derived from the animal body itself.

He made many attempts to determine which of the two kinds of electricity was produced in the nerves, and which in the muscles; thinking that one of these tissues would naturally be positive and the other negative. But he finally considered the electricity to be distributed on the exterior and in the interior of the muscle very much as it is on the outer and inner surfaces of a Leyden jar. The nerve, with its ramifications, leading from the interior of the muscle, he regarded as analogous to the chain and knob of the Leyden jar; and when it was connected by a metallic arc with the exterior of the muscle, the animal electric battery was discharged, and a convulsion ensued from the stimulus of the shock.

In this way Galvani accounted for the phenomena witnessed in his experiments. He was far from believing that he fully understood all their relations; but he was convinced of their importance, and had no doubt that their obscurities would disappear on further investigation. He especially noticed the advantage of uniting in the armature of the limbs, or in the connecting arc, different metallic substances. "There is an additional peculiarity," he says, "which deserves attention, and which I have very often observed in regard to the connecting arcs; namely, that they are vastly more efficient when composed of different metallic substances than when consisting of one and the same metal." But he does not undertake to explain this peculiarity; although he found, that, of various metallic combinations, some were more effective than others.

He was perfectly clear as to the prime importance of what he had actually discovered, and the secondary value of what was only inferential; and this distinction is expressly stated in his opening chapter. "From all these particulars," he says, "investigated and established by a long series of experiments, we have not only shown that these contractions are due to electricity, but have furthermore been able to indicate certain conditions and laws by which they are regulated. . . . We have also appended to the narrative a number of corollaries, together with some additional conjectures and hypotheses, mainly in the hope of opening a way for further experiments, by which we may at least be enabled to approximate the truth, even if we cannot expect to reach it altogether."

There are some curious surmises in Galvani's book as to the possible future value of his discoveries in the treatment and cure of nervous disease. He recognizes the difficulty of their application to pathology, on his own hypothesis of an animal electricity as the cause of muscular contraction. But they are interesting in connection with the imperfect state of pathological knowledge in his day as compared with the present, and some of his remarks sound almost prophetic. "In regard," he says, "to the cure of paralysis, I see that it is a matter of great uncertainty; for it is not easy to say whether the disease be caused by a degeneration of structure in the nerves and brain, or whether it be due to an obstruction of the supposed electrical circuits by the deposit of some non-conducting material. But perhaps," he adds, "the whole thing will some day or other be cleared up by further practice and experience."

Galvani did not have to wait long to see a very important advance made in the path which he had opened. One of his contemporaries was Volta, who had already been for fifteen years professor of physics in the University of Pavia, and who was admirably fitted, both by his capacity and his attainments, for treating a new subject with success. Volta was greatly attracted by the novelty and character of Galvani's experiments. He repeated them, with many extensions and variations; and he not only verified their results, but was enabled to throw a new light on the immediate cause of their phenomena. His first investigations were communicated in the form of two letters to the Royal Society of London in 1793, under the title "An Account of some Discoveries made by M. Galvani, with Experiments and Observations on Them."[1] In this communication he refers to Galvani's treatise, published two years before, as containing "one of the most splendid and striking discoveries, as well as the germ of others in addition."

Volta naturally approached the subject rather from its

[1] Philosophical Transactions of the Royal Society of London for the Year 1793, p. 10.

physical than its physiological side, and he had a certain mathematical habit of mind that led him to appreciate the value of quantities in such an investigation. He began by determining, in an approximate way, the amount of electrical stimulus capable of producing convulsions in the muscles of the frog. By using successively smaller and smaller charges from a Leyden jar, he found that the entire and living frog might be convulsed by electric discharges which would give but very faint sparks, and which would affect only the most delicate of the electrometers then in use. But a recently killed frog, prepared after the method of Galvani, was more sensitive still. In this condition it could be made to contract by an electric discharge fifty or sixty times more feeble, and quite imperceptible with any electrometer, unless by aid of the condenser. In this way he exhibited Galvani's frog as a new kind of electrometer, or electroscope, of extraordinary sensibility; and he named it, accordingly, the "animal electrometer." This showed conclusively that in all the experiments with induced electricity the dissected frog might be convulsed by an electric discharge too feeble to be detected by other means.

Volta then began to investigate the source of electricity in Galvani's third set of experiments; namely, those with the dissected frog and the metallic arc. Galvani, it will be remembered, attributed this electricity to the different condition of the animal tissues, and always made his connection between the nerve on the one hand and the muscle on the other, as he would between the two surfaces of a Leyden jar. With the connection made in this way the effect is really more striking, because the whole of the galvanic current passes through the isolated nerve, and thus produces a corresponding excitement. But Volta found that the two separate elements of nerve and muscle respectively were not essential to the result. If one metallic armature were placed on the muscle, and another also on the muscle near by, or if they were placed on corresponding muscular parts of the two legs, a contraction ensued on connecting them by the metallic arc. The same thing happened with the nerve. If an armature of silver were placed on one part of the crural nerve above its junction with the muscles, and one of tin-foil at another part of the same nerve still higher up, and connection made between them, the leg was convulsed, notwithstanding that the whole of the muscles, as well as a portion of the nerve, were outside the electric circuit. This did not correspond with the idea of an animal electric apparatus consisting of the nerve and muscle combined. It was necessary to look elsewhere for the source of the electricity under these conditions; and Volta found it in the contact of dissimilar metals. This contact developed but a minute quantity of electricity, too small to be recognized by any other means at the command of the experimenter; but the dissected frog was an electroscope of such sensibility, that it responded to the test, and betrayed the electric discharge by a convulsion.

Volta was in very much the same position for this part of the discovery as Galvani had been in regard to his own earlier observations. He was compelled to recognize its reality and importance without being able to give its explanation. "I confess," he says, "it is not easy to understand how or why the application of dissimilar armatures — that is, of different metals — to two similar parts of the animal, or even to neighboring parts of the same muscle, should disturb the equilibrium of the electric fluid, and drive it from its condition of repose into one of active and continued displacement. But whatever may be the cause of the phenomenon, and whether intelligible or not, it is nevertheless a fact abundantly established by the experiments already detailed, and further corroborated by those which follow."

For several years Volta continued the study of this subject, and at last found a way of greatly increasing the intensity of the phenomena. This was by multiplying the number of pairs of his dissimilar metals so as to form the Voltaic pile. His results were given in 1800 in a letter to the Royal Society,[1] containing a description of his new contrivance. Each member of the pile consisted of a plate of zinc in contact with a plate of silver. This was covered by a layer of moistened paper or membrane. Then followed a second pair of zinc and silver plates, and so on, alternately; the two metals always recurring in the same order, and each pair being separated from those above and below by moist membranes. With this arrangement the intensity of the action, on connecting the extremities of the pile, was sufficient to cause shocks and muscular contractions in the hands and arms, and to affect the senses of sight, hearing, taste, and touch. Its operation was compared to that of the torpedo and electrical eel, and it received the name of the "artificial electrical organ."

The peculiarity of this discovery was, that it introduced into electrical science two new features. Before that time, electricity had been obtained only by the combined use of different substances, one of which must be a conductor, the other a non-conductor, as in the frictional machine, the Leyden jar, and the electrophorus; and, when an

[1] Philosophical Transactions of the Royal Society of London for the Year 1800, p. 403.

electrified body was touched by a conductor, the whole of its charge was expended, and it could be electrified again only by receiving a new charge from an extraneous source. But in the electric pile Volta found, to his surprise, a source of electricity in substances all of which were conductors, and, furthermore, a source which seemed inexhaustible; because, however often the two ends of the pile were touched, the same shock was repeated with the same intensity. The inherent force of the pile seemed to urge the electric fluid incessantly forward, returning upon itself in a continuous flow so long as the circuit of conductors remained complete.

Volta attributed this action entirely to the contact of dissimilar metals. This view is plainly expressed in the title of his letter to the Royal Society, and is repeatedly insisted on in the course of the communication. The fluids or moistened membranes interposed between the metallic pairs he regarded only as conductors; and, when a saline solution was found to be more efficient than water alone, he thought it was because the salt increased the conducting power of the water. In reality, the source of power in the Voltaic pile is the chemical action between the fluid and the metals, one of which is more oxidizable than the other; and, for every unit of electric force sent through the circuit, a definite quantity of material undergoes chemical transformation. The contact of dissimilar metals is not essential to the result; for a current may be produced without it, if we use two different liquids, of which one is more decomposable than the other.[1]

Neither the first nor the second explanation of the phenomena observed was therefore entirely right, as we

[1] Daniell: Introduction to the Study of Chemical Philosophy. London, 1843, p. 492.

understand it. Volta imagined the metals in his pile to act by contact alone, while they are really consumed by oxidation. In Galvani's experiment, the connecting arc where the two metals join is, as Galvani supposed, a conductor, and nothing more; but its different extremities, when applied to the animal tissues, form with them a galvanic battery of a single cell, and thus produce the current which excites their contraction. We now use it for physiological purposes in a different way. We place the two metallic plates in a cup of acidulated fluid, and make the returning current pass through the muscles and nerve. But the effect is the same, for the active force of the current is equal in both directions.

The results obtained by these two investigators were distinct from each other, but of almost equal importance. Galvani discovered the electric action on nerves and muscles which bears his name. Volta gave to science a new apparatus, by which current electricity is transmitted in a continuous circuit. Each investigator was partially at fault in the theoretical explanation of his own discovery. But the value of the discovery has remained, and has even largely increased in the course of a century, notwithstanding the difference in its interpretation. It also appears, from the history of the circumstances, that the second of these discoveries was a consequence of the first; and nothing can show more clearly the unbroken connection of events in the progress of science, which may sometimes extend to the most unexpected ramifications. It is plain that we should not be to-day in possession of the electric light, were it not for Volta's discovery of current electricity; and Volta produced his electric pile in trying to investigate the contraction of Galvani's frogs.

During the early part of the present century, experimenters were busily occupied with galvanic electricity. They studied its action in a great variety of ways, and thus became acquainted with its mode of operation, and able to appreciate its phenomena. They found that when it was transmitted through any portion of a muscular nerve it produced contraction at the instant of closing the circuit; that afterward, while flowing through the nerve in a uniform current, it was without effect; and that contraction again occurred when the current was discontinued. Even during the passage of the current, any sudden variation in its intensity, either of increase or diminution, would produce the same effect. It was the change in the electrical state of the parts, rather than their actual condition, which operated as a stimulus and provoked the contractions. The direction of the current, the length of nerve included within the circuit, and the frequency of making and breaking the connection, — all had a certain influence in the result. Experimenters found themselves provided with a new agent, by which they could investigate nervous action at their leisure, in animals deprived of life, with the nerves and muscles exposed to view, and with far more delicacy and certainty than ever before; and the familiarity thus acquired with the operation of the galvanic stimulus prepared them for its more effective use in future investigations.

The next event of sufficient consequence to form an epoch in the history of nervous physiology was the discovery, in 1822, of the distinction of motor and sensitive properties in the two roots of the spinal nerves. Eleven years before, Sir Charles Bell [1] had made some experiments on the spinal nerve roots of a recently killed rabbit,

[1] Idea of a New Anatomy of the Brain; submitted for the Observation of his Friends. By Charles Bell, F.R.S.E.

in which it appeared that mechanical irritation of the anterior roots caused convulsion in the corresponding muscles, while a similar irritation of the posterior roots had no such effect. This, however, was not supposed to show an anatomical separation between the two powers of motion and sensibility, but between those of volition and consciousness on the one hand, and involuntary nervous action on the other. The ideas of the author on this subject were not accepted by his colleagues. In fact, they were not really published, but only printed in pamphlet form for private distribution. They hardly attracted the notice of the profession, and had no influence on the medical doctrines of the day.

Subsequently, in 1821, Bell published [1] the remarkable observation that if the seventh cranial nerve, distributed to the face, were divided in a living animal, the movements of facial expression were abolished; while the faculty of sensation could be destroyed by cutting the corresponding branches of the fifth pair. Thus the two main properties of nervous endowment, generally associated with each other, occupied in this instance distinct situations; since the seventh was shown to be a nerve of motion only, the sensibility of the face being supplied from another source. This fact was connected, in the researches of Sir Charles Bell, with theoretical considerations of a different bearing; but it had evidently a great importance of its own, and was received on all sides with much interest. It especially attracted the attention of Magendie, then at the height of his activity in physiological investigation. Magendie repeated Bell's experiments on the seventh cranial nerve, and verified their results as to its exclusively motor properties. But he was

[1] Philosophical Transactions of the Royal Society. London, 1821, p. 398.

also intent on exploring various other parts of the nervous system; and in 1822 he performed and published his experiments on the spinal nerve roots.[1] His description is a model of directness and simplicity, free from any unwarranted inference or assumption. "I had long been wishing," he says, "to try the experiment of dividing the posterior roots of the spinal nerves, but had never succeeded in doing so, owing to the difficulty of opening the vertebral canal without wounding the spinal cord, and inflicting severe or fatal injury upon the animal. But a month ago I received at the laboratory a litter of pups only six weeks old; and it seemed a good opportunity for another attempt at the operation. This time, in fact, I succeeded, by the aid of a very sharp scalpel, in laying bare, almost at a single stroke, the posterior half of the spinal cord with its membranes. I then had no difficulty, after opening the dura mater, in bringing into view the posterior roots of the lumbar and sacral nerves; and, by lifting these roots on the blades of a pair of fine scissors, I was able to divide them on one side without injury to the spinal cord. Not knowing what would be the effect of this operation, I closed the wound by a suture in the integument, and kept watch of the animal. At first it looked as if the limb on that side were entirely paralyzed. It was insensible to punctures and to pressure, and also appeared motionless; but soon afterward I was surprised by seeing it move very distinctly, although it was completely and permanently insensible. After trying a second and a third experiment with exactly the same result, I began to think it probable that the posterior roots of the spinal nerves might have a function distinct from that of

[1] Journal de Physiologie Expérimentale et Pathologique (Paris, 1822), tome ii. p. 276.

the anterior roots, and more especially devoted to sensibility.

"I next thought of dividing the anterior roots, leaving the posterior roots entire; but that was a thing more easily said than done. There appeared to be no way of getting at the anterior surface of the cord without involving the posterior roots; and at first it seemed a hopeless undertaking. After considering the matter for a day or two, I tried to reach the anterior roots with a sort of narrow-bladed cataract-knife, by passing it in front of the posterior roots, and then turning its cutting edge forward against the bodies of the vertebræ. This plan failed, owing to the impossibility of avoiding hemorrhage from the large veins in the vertebral canal; but, while making the attempt, I found, that, by drawing aside the *dura mater*, I could catch a glimpse of the anterior roots just where they are about to pass through the investing sheath. This was enough; and in a few seconds I had divided as many of them as I wished, on one side only. It may be imagined with what curiosity I awaited the result. It finally came in a perfectly unequivocal form; for the limb, though retaining its sensibility, was completely relaxed and motionless. Finally, not to leave any thing undone, I divided both the anterior and posterior roots, causing entire loss of sensibility and motion."

Magendie began his experiments, therefore, by dividing the nerve-roots, and thus abolishing their functions. In pursuing the subject, he tried the effect of their irritation, employing for this purpose both mechanical agents and the stimulus of galvanism. The result of these experiments corresponded in general with those obtained by the method of division; that is, galvanization of the anterior roots caused muscular contraction, and that of the posterior roots sensation. There were also certain other phenomena which did not altogether coincide with the doctrine of a complete separation between the two functions, and which were not fully understood until some years later. But Magendie took cognizance of every experimental fact, whether he understood it or not; and his own subsequent researches on "recurrent sensibility" explained many of the irregularities which he at first encountered.

This distinction between the spinal nerve roots was of greater importance, because it indicated a general plan of arrangement for the nervous system throughout the body. It immediately became a matter of criticism and verification for the leading physiologists of Europe; and the result was a complete acceptance of Magendie's discovery. In Germany, Johann Müller, then professor of anatomy and physiology in the University of Bonn, endeavored to avoid the difficulties of so serious an operation on warm-blooded animals by experimenting on frogs. He examined the spinal nerve roots in these animals, both by the method of section, by mechanical irritation, and by galvanism.[1]

"These experiments," he says, "have been rewarded with the most brilliant success. They are so easy of application, so sure and so decisive, that any investigator may now readily satisfy himself of one of the most important truths in physiology. The phenomena are so regular and satisfactory, that, for simplicity and certainty of results, they fully compare with any crucial experiment in the physical sciences. Galvanization of the separated anterior roots at once causes active convulsions; that of the posterior roots never gives rise to any sign of spasmodic action."

[1] Handbuch der Physiologie des Menschen. Coblentz, 1837. Band i. p. 651.

In this way the distinct endowment of the two kinds of nerve-fibres was experimentally established. Once placed on this footing, the pursuit of nervous physiology was greatly increased in efficiency and extent. By applying the galvanic stimulus to a spinal nerve above or below the point of section, its mode of action was determined by the excitability of its motor or sensitive fibres. The same method was employed for the cranial nerves, both externally and at their roots; and every branch of inosculation was scrutinized by the same means. It is hardly possible to overestimate the change thus introduced into the study of the nervous system, and the facilities which it supplied for further investigation.

We now come to Marshall Hall's discovery of the reflex action of the spinal cord. Up to this time, the motor and sensitive properties of the nervous system had been understood by physiologists as almost wholly subservient to voluntary motion and conscious sensibility. These functions resided in the brain, as the organ of intelligence and the source of all spontaneous action. From it the motor nerves transmitted to the muscles the commands of the will, while the sensitive nerves brought to it from without the impressions made on the integument. The spinal cord was part of this apparatus of transmission. Like the spinal nerve roots, under the influence of galvanism, its posterior columns were found to be sensitive, and its anterior columns excitable; and its complete division at any one point abolished voluntary movement and sensation in the parts below. It was the channel through which the spinal nerves held their connection with the brain.

But Hall observed, that, notwithstanding the loss of sensation and volition after removal of the brain, the animal might still be capable of motion in the limbs, provided the spinal cord remained. Although phenomena of this kind had already been noticed in several instances, they had not been demonstrated with sufficient distinctness to fix the attention of physiologists. Hall's observations were first announced in a communication to the London Zoölogical Society in 1832. They were further embodied in his "Lectures on the Nervous System and its Diseases," in 1836, and still more formally presented in his "Memoirs on the Nervous System," in 1837. Their simplest demonstration was given by him in the following way: In a living and uninjured frog the signs of sensation and volition are manifested whenever the skin is irritated at any point, since the animal feels the impression, and responds to it by a voluntary motion. When the head is cut off, or the spinal cord divided at its upper part, the limbs are paralyzed, and sensation is abolished. But if, while the animal is in this condition, one of the feet be pinched, the limb is drawn upward; and the movement may be repeated as often as the irritation is applied. Such a movement is very different from that caused by galvanizing a motor nerve; since the stimulus in this instance is applied to the skin, and the muscles react in consequence. Both skin and muscles must retain their connection with the spinal cord; since, if this connection be cut off, no movement takes place on pinching the foot. Finally, the nerves remaining uninjured, if the spinal cord be broken up, all reaction ceases, and irritation of the skin has no further effect. But does not such a violence destroy the physiological property of the muscles, and in that way prevent their contraction? This doubt is removed by applying a galvanic current to the muscles themselves, when they are at once convulsed, showing that their contractile power is unimpaired.[1]

[1] Lectures on the Nervous System and its Diseases. London, 1836, p. 19.

Thus, it is the spinal cord which acts, independently of the brain, as a medium of communication between the integument and the muscles. The stimulus conveyed inward, through the sensitive nerves to the cord, is thence reflected outward through the motor nerves to the muscles. From this circumstance it received its name of "reflex action;" and, since it was first studied and demonstrated in the spinal cord, it was generally known as the "reflex action of the spinal cord."

But the same form of activity was afterward found to be very widely extended in the nervous system. The medulla oblongata has its own centres of reflex action, either directly or indirectly essential to the continuance of life. Wherever there is a ganglionic mass of nervous matter, with motor and sensitive fibres originating from it, there is a similar focus of nervous power, often quite disconnected with consciousness and volition. In a state of absolute insensibility, in man or animals, a touch upon the cornea will cause closure of the eyelids, irritation of the anus will increase the contraction of the sphincter, and the contact of a solid body with the fauces will excite the movement of deglutition; and in all these instances the reaction disappears when its special nervous centre is destroyed. Similar facts were soon observed in man in cases of paralysis, as where movements are produced in paraplegic limbs without the knowledge of the patient; and nearly all the phenomena of convulsive affections were seen to have their origin in some unusual irritation of a nervous centre, or in the morbid exaggeration of its excitability. From that time forward the reflex action of the nervous system entered more or less into the whole study of its normal and diseased conditions.

The next topic of special interest in this connection is

the influence of the nervous system on the organs of circulation. The earlier approach to definite knowledge on this subject was a very slow one. It consisted mainly in establishing the fact of contractility in the arteries, and was accomplished by experimental inquiries extending over a long time, — from those of Hunter, on the arteries of animals, in 1793, to those of Kölliker,[1] in 1849, on the constriction caused by galvanism in the popliteal and tibial arteries of an amputated human limb. A new epoch in the physiology of the circulation was reached in 1851, when Claude Bernard[2] published the discovery that division of the sympathetic nerve in the neck is followed by enlargement of the blood-vessels on the corresponding side of the head. This effect is so striking and so constant, that, when once announced, there was no difficulty in its verification, and no doubt as to its reality. Almost immediately after the section of the nerve, an increased vascularity becomes visible in the conjunctiva, the mucous membrane of the nostril, lip, tongue, and cheek, and in all parts of the skin on the affected side. It is most distinctly seen in the ear of the white rabbit, because the organ presents a thin expansion of semi-transparent tissue, convenient for observation, and because the two ears, placed side by side, afford a ready criterion of any change in vascularity. In the first report of these experiments, the attention of the observer was principally directed to the local increase of temperature, which also follows division of the sympathetic; but this was afterward seen to depend on the greater activity of the circulation, which was then recognized as the primary and characteristic result of the operation.

---

[1] Zeitschrift für wissenschaftliche Zoologie. Leipzig, 1849. Band i. p. 259.
[2] Comptes rendus de la Société de Biologie. Paris, année 1851. Tome iii. p. 163.

This fact, of enlargement of the blood-vessels from division of the sympathetic nerve, at once excited a lively interest among physiologists. Hardly a year had elapsed, when a second observation, equally important with the first, was made almost simultaneously by Brown-Séquard in Philadelphia,[1] Bernard in Paris,[2] and Waller in London;[3] namely, that the condition of the circulation on that side of the head where the sympathetic has been divided may be regulated at will by experimental means. Suppose that increased vascularity has been produced by division of the sympathetic in the neck: if the stimulus of galvanism be now applied to the divided nerve above its point of section, all the previous results of the operation disappear. The blood-vessels contract, the volume of the circulation diminishes, the local temperature is reduced, and the parts resume their normal color, or even become more pallid than before. When the galvanization is suspended, the former conditions return, with all the accompanying phenomena of vascularity, temperature, and redness; and the circulation in the part may be in this way alternately increased and diminished for many successive repetitions of the experiment.

It thus appears that the muscular coat of the arteries, supplied with nerve-fibres from the sympathetic, is influenced by them in nearly the same way as the voluntary muscles are controlled by the cerebro-spinal nerves. Division of the sympathetic paralyzes the involuntary muscular fibres, relaxes the arterial walls, and allows a larger quantity of blood to pass through the vessels of the part. Galvanization of the nerve, on the other hand, stimulates the muscular fibres to contraction, narrows the caliber of the

[1] Philadelphia Medical Examiner. 1852. Vol. viii. p. 489.
[2] Comptes rendus de la Société de Biologie. Paris, 1852. Tome iv p. 168.
[3] Comptes rendus de l'Académie des Sciences. Paris, 1853. Tome xxxvi. p. 378.

vessels, and so reduces the volume of the circulating blood. The knowledge of these facts introduced into the nomenclature of the nervous system a new title. There were evidently nerve-fibres which acted upon the blood-vessels to call into operation their contractile power; and the nerves possessing such a function, then known for the first time, naturally received the name of the "vaso-motor nerves."

But, in following out this subject, an observation was soon met with of very unexpected character; namely, that certain other nerves, on being stimulated, instead of producing contraction of the blood-vessels, caused their relaxation, and thus increased the activity of the circulation. This was first shown by Bernard in the case of the submaxillary gland. This organ is supplied with sympathetic fibres from the superior cervical ganglion and the carotid plexus, and with cerebro-spinal fibres from the lingual nerve and the chorda tympani. Galvanization of its sympathetic filaments produces, as in other similar instances, contraction of the vessels, and a diminished blood-supply. But if the stimulus be applied to the lingual nerve above the situation of the gland, or to the chorda tympani, which unites with it, the result is exactly the contrary: the blood-vessels enlarge and the circulation is more active so long as the galvanization continues; and this effect is equally marked if the nerve be divided, and galvanized between its point of section and the gland.

A similar influence was found to reside in other parts of the nervous system; and the nerves possessing the power of thus causing vascular enlargement were called "dilator nerves." No sooner was this fact established in a general way, than it served to explain a singular phenomenon which had been noticed many years before, but which had thus far been regarded as exceptional; namely, the influence of the pneumogastric nerve on the action of the heart. As a general rule, if the nerve going to a muscular organ be divided, the muscle is paralyzed; and, if the nerve be stimulated, there is muscular contraction. As the heart receives filaments from the pneumogastric nerve, we should naturally expect that its action would be diminished by section of this nerve, and increased by its stimulation. But the effect is really the reverse. If the poles of a galvanic apparatus be applied to the pneumogastric nerve in the neck, the cardiac pulsations are reduced in frequency; and, when the strength of the current is increased to a certain degree, they stop altogether. The heart lies quiescent, in a state of relaxation, its movements remaining in abeyance while the galvanization goes on; and, when it is suspended, they recommence with undiminished energy. The influence exerted in this case is not reflex, but direct, in its operation: for if the nerve be divided, and galvanized above its point of section, there is no result; but, if the stimulus be applied below the section, its retarding action on the heart is at once manifest. Furthermore, the power of this nerve to restrain the cardiac movements, like the motor influence of a spinal nerve, is limited in duration. You cannot permanently arrest the heart, and so kill the animal, by continued galvanization of the pneumogastric. When the galvanization of the nerve has been kept up for a certain time, the heart begins to beat again. Its pulsations recur at first slowly, afterward more frequently; and at last they are restored in full regularity, notwithstanding the continuance of the galvanic current. The nerve has lost its power by exhaustion, and cannot again manifest its controlling force, unless allowed to recover by repose. But the heart is still sensitive to the same influence; and, if the electrodes be shifted to the pneumogastric of the opposite side, it stops as quickly as before.

It must be admitted, therefore, that the influence of the pneumogastric, whatever it may be, which controls the heart's movement, is transmitted from within outward to the peripheral extremities of the nerve; and that it acts in a positive manner, though producing a negative result. It is the most striking illustration of a kind of action in the nervous system unlike any of those formerly known, but not the less real for being difficult to understand. This is the so-called "action of arrest," an influence which passes through a nerve from its origin to a muscle, and by which the muscular contraction is suspended. As often happens in such cases, when the existence of this mode of action was once realized, it appeared that there were other instances of the same thing which had been overlooked. All the sphincter muscles, though habitually in a state of involuntary contraction, are suddenly relaxed at certain periods by an influence coming from within. The blood-vessels generally receive both kinds of nervous impression; and, by the varying preponderance of one or the other, they are alternately made to contract or dilate, with all the accompanying changes of local circulation. In this way it became possible to explain the mechanism of temporary physiological congestions, as in secreting glands, or in the alimentary canal during digestion; and those of longer continuance, with increased nutrition, like the growth of the uterus and mammary glands during pregnancy, as well as morbid disturbances of the circulation in disease.

Quite a different line of investigation was inaugurated by Helmholtz, in 1851, for determining the rapidity with which nervous action is transmitted through the motor nerves.[1] The idea of measuring, with any approach to

[1] Comptes rendus de l'Académie des Sciences. Paris, 1851. Tome xxxiii. p. 262.

numerical precision, the movement of the intangible nerve-force through its fibres, would seem at first almost beyond the scope of reality; and yet it was accomplished with satisfactory success, and by perfectly genuine experimental methods. It would have been impossible, were it not for improvements in the galvanic apparatus and the registering machines, which have played so important a part in the more recent investigation of animal physics. The knowledge and use of induced electric currents we owe to Faraday; and their discovery in 1831, with all the related phenomena of electro-magnetism, magneto-electricity, and the production of instantaneous and rapidly alternating opposite currents, practically revolutionized the use of electricity for medical purposes. It also supplied especial facilities for experiments on the rate of transmission of the nerve-force.

The conditions necessary for such an experiment were twofold: first, an instantaneous induced current, for causing a single muscular spasm; and, secondly, an automatic registering apparatus, which should mark the exact time, both of the electric stimulus and of the muscular action. By this means a definite result was obtained. With the electrodes applied to the muscles of a frog's leg, an interval amounting to the one-hundredth part of a second appeared between the closure of the circuit and the contraction of the muscle. The muscular contraction, therefore, was not an instantaneous effect, but required a certain time to get under way after the application of the stimulus. This interval was not perceptibly altered on applying the electrodes to the nerve at or near its entrance into the muscle; but, if they were applied higher up on the nervous trunk, the delay became longer; and it was increased, in subsequent trials, exactly in proportion to the distance between the muscle and the point of nerve-stimulation. It represented, accordingly, the time required for the nerve-force to traverse a given length of nerve-fibre; and, by repeating the test under various conditions, its rate of movement was fairly determined.

These experiments, first performed on the nerves and muscles of the separated frog's leg, were afterward extended to those of the living man, the electric stimulus being applied to the skin over the situation of a nerve at different points, and the contraction indicated by the swelling of the parts over the muscle. In all the investigations thus far, the nerve was excited by the artificial stimulus of electricity. Subsequently, Burckhardt[1] simplified the experiment, and increased the value of its results, by employing, instead of electricity, the natural stimulus of volition. The subject being placed in connection with a proper registering apparatus, the signal for voluntary effort was given by the sound of a bell, and the movement was performed in different instances by different muscles. Under these conditions, the time necessary, both for volition and for the mechanism of muscular contraction, would be the same in all cases; but that required for traversing the motor nerve would vary according to the muscle employed. A voluntary impulse starting from the brain would arrive at the deltoid muscle after travelling a certain distance; but it would follow a longer route to reach the adductor of the thumb. The distribution of the crural nerve to the quadriceps extensor muscle is comparatively remote from its origin, and that of the sciatic nerve to the dorsal muscles of the foot more distant still. The difference in the time of muscular contraction observed in these cases corresponded with the different lengths of nerve conveying

[1] Die physiologische Diagnostik der Nervenkrankheiten. Leipzig, 1875, p. 32.

the stimulus; and it gave for the passage of the voluntary impulse in man through the motor nerves an average rate of twenty-seven metres per second.

A similar contrivance for measuring the transmission of tactile impressions through the sensitive nerves gave their rate of movement as forty-seven metres per second, showing a greater rapidity of transmission for sensitive impressions than for motor impulses; and, by comparing the result of further experiments with the known quantities obtained in this way, a close estimate was reached of the time needed for the operation of the different senses, and even for the cerebral acts of perception and will.

When these observations were made on different persons, there was always a certain amount of variation in the result; the nervous action being in some cases more rapid, in others comparatively slow. This brought within the range of definite physiological experiment a fact first noticed in astronomical observatories; namely, that two observers, both watching for the same event, seldom see and record it at the same time. There is a difference in the quickness with which they receive its impression on the senses, and in each case there is a certain amount of delay; so that, in neither the one nor the other, is the phenomenon perceived at the time of its actual occurrence. In astronomical operations where extreme accuracy is required, as in transit observations for the determination of longitude, this personal imperfection of the observer needs to be corrected from a previous examination of his habitual error. In the report of the United States Geographical Surveys for 1877, it is stated by Dr. Kampf that the personal error from this source varies somewhat in the same individual from day to day; so that its amount should be ascertained, and the proper correction made for each person, whenever a longitude observation is to be taken.

There is still another point of interest in connection with the modern study of the nervous system; namely, the localization of function in different regions of the brain.

The most striking part of this subject relates to the special centres for motion and sensation in the cerebral convolutions. There are few discoveries which have seemed more at variance with our former convictions than that of the existence of these centres. Both the substance and the surface of the cerebrum had often been subjected to experimental examination, both in the living and the recently killed animal, without showing any signs of muscular reaction; and it was the universal belief among physiologists, that none of its parts were directly subservient to any form of motion or sensibility. So industrious and skilful an observer as Longet [1] declared, in 1869, that he had "irritated by mechanical means the white substance of the hemispheres in dogs, cats, rabbits, and birds, and had stimulated it by the application of potassa, nitric acid, or the actual cautery, as well as by the passage of electric currents in various directions, without ever bringing into play the involuntary muscular contractility or convulsive movements; and similar applications to the gray substance of the convolutions were equally without effect."

These failures made it seem hopeless to anticipate any further result from direct exploration of the cerebral substance. The brain was generally regarded as so exclusively the organ of intelligence, that it could not be expected to respond to the irritation of physical agencies. In the extremely condensed and valuable work of Ecker on the convolutions of the brain, the author says in his preface, that "the accurate observation of patients by their physi-

---

[1] Traité de Physiologie, 3me édition, Paris, 1869. Tome iii. p. 146.

cians, in connection with careful autopsies, is the only means by which we can ever hope to learn the physiological significance of particular cerebral convolutions."

This preface was dated March, 1869; and in 1870 Fritsch and Hitzig [1] discovered, by experiments with galvanism on the dog's brain, that there are certain parts of the cerebral convolutions where this stimulus always produces definite and unmistakable movements on the opposite side of the body. The contraction of certain groups of muscles, and consequently particular movements in the trunk or limbs, are connected with the stimulation of particular points of the brain; and, when such a point is once found, the corresponding movement may be reproduced at will by repeating the application of the stimulus. There is plainly, in some way or other, a communication, through definite nervous routes, from the special centre of motion on the surface of the hemisphere to the motor tract in the medulla and spinal cord, and thence to the muscle which performs the contraction. In all the animals used for experiment, these centres are grouped in certain regions, while other portions of the cerebral surface show no similar indications; and by comparing their position in different species, aided by observations in human pathology, it appears that in man the motor centres for the body and limbs of the opposite side are mainly located in the anterior and posterior central convolutions, immediately bordering on the fissure of Rolando.

Thus the earlier failures and the more recent success in the discovery of the motor centres are both explained. It is true that a large part of the cerebral surface is unexcitable by artificial means. You may apply the galvanic

[1] Archiv für Anatomie, Physiologie, und wissenschaftliche Medicin. Leipzig, 1870, p. 300.

electrodes to twenty different points of the convolutions without the least sign of a muscular contraction. But on the twenty-first trial you may strike one of these centres, and then the muscular spasm immediately follows. We are now so well acquainted with their probable location in any particular brain, that we need not lose a great deal of time in finding one of them; but, before the geography of their distribution was known, it is not surprising that experimenters should have overlooked their existence.

The account given of their discoveries by Fritsch and Hitzig was exceedingly well expressed, and bore internal evidence of its faithfulness of description. All the details of experimental procedure were fully explained; and the results were stated in such a way that other observers could easily follow in the same direction, and test their reality. Some doubt has been entertained in various quarters as to the interpretation of the phenomena, and particularly how far the muscular contractions might be due to a diffusion of the galvanic current beyond the limits of its immediate locality. But this doubt was removed after repeating the experiments by a variety of methods; and the existence of the motor centres was corroborated by subsequent discoveries in the minute anatomy of the parts, and the local alteration of structure in cases of hemiplegia. The subject is still so new, and in so active a condition of development, that it can hardly be presented in the form of a complete or well defined physiological doctrine. But it is evidently a matter of great importance, and is probably receiving at this time a larger share of attention than any other single topic relating to the nervous system.

Its latest extension is connected with the centres of *sensation* in the cerebral hemispheres. As some parts of the convoluted surface of the brain are plainly subservient to muscular action, it is natural to infer that the remaining unexcitable portions may have a similar connection with sensibility; and it is asserted by some, with more or less confidence, that the senses of touch, taste, smell, sight, and hearing are separately located in as many different regions of the cerebral cortex. The experimental evidence of these localizations is far from being altogether satisfactory; but in one instance, at least, it is very striking, and indicates beyond doubt a close relation of visual sensibility with the "angular convolution" on the posterior and lateral part of the cerebral hemisphere. If this convolution be extirpated, the operation is followed by blindness of the opposite eye, without any other perceptible disturbance of either motion or sensibility. As other parts of the brain-surface, of equal or greater extent, may be removed without causing impairment of vision, it is difficult to avoid the conclusion that this region has a special connection with the sense of sight. The exact nature of the connection will doubtless be better understood from further investigation.

The complete history of physiological science for the last hundred years, in regard to the nervous system, can hardly be given in the space of a single lecture. But its most important advances are so connected with each other, that they have a relation very much like that of cause and effect. When a new subject of inquiry is first opened, the progress is for some time a slow one. There are difficulties in the way, which must be overcome by repeated experiments, by gradual improvement in the apparatus, and by better methods of procedure. The causes of the phenomena are imperfectly understood, and their relation with other observed facts is not immediately apparent.

But when the knowledge acquired has reached a certain point, its advance becomes more rapid. Every addition enlarges the circle of its operations, and enables it to execute them with greater facility. And the results attained by this means are not always those most directly anticipated. One discovery often leads to another by bringing into view incidental facts, which, in turn, become the sources of new information; and in that way it creates opportunities for future progress which are sometimes realized in their fullest extent only after an interval of several generations.

There is no more interesting department of physiological study at the present day than that of the vaso-motor nervous system. Since it was first practically inaugurated by Bernard, thirty years ago, it has been cultivated by many observers, and has received a wide extension in many directions. It touches on most important points of pathology, as well as the functions of health. The connection between secretion and blood-supply, the mechanism of congestions, the dependence of external disturbances of the circulation on disease of internal parts, the red cheeks of pneumonia, the hectic of pulmonary phthisis, and the existence of nervous centres in the cerebro-spinal system, where these changes are controlled by reflex action, are all made capable of investigation by knowledge which has been derived from this source. But how could their study have been even attempted, unless we were already in possession of the simpler facts of reflex action of the spinal cord, and the different behavior of motor and sensitive nerve-fibres? All the variations in the effect produced by electric stimulus of different kinds and intensity, the comparative influence of direct and inverse currents, the exhaustion of nerves by continued stimulation, and their recovery by repose, together with many other similar conditions, were indispensable stages in the progress of discovery, and were the fruit of many intermediate investigations. But each series depended, for the possibility of its existence, on another which had gone before; and they all had their origin, in a continuous line of descent, from the experiments of 1789, in Galvani's laboratory at Bologna.

LECTURE II.

BUFFON AND BONNET IN THE EIGHTEENTH CENTURY.

Mr. President and Gentlemen, — I shall ask your attention in this evening's lecture to two remarkable phases of physiological doctrine which once held a conspicuous place in the world of science; namely, Buffon's theory of Organic Molecules, and Bonnet's theory of the Inclusion of Germs. Notwithstanding that both are now so obsolete that perhaps the majority of my hearers will hardly recall their meaning, they were formerly considered of sufficient importance to be test questions in the controversies of the period, and they are fair examples of the fluctuating estimate to which such subjects are liable at different times.

It sometimes happens, that, in searching the annals of medicine, we find allusions to systems and theories which have so far gone out of vogue that we know them only by name. They are no longer part of our scientific doctrines; and they have so little present value or significance, that we seldom spend the time even to weigh their merits or defects. And yet these systems once had their day. They were prominent topics of discussion and interest; and they divided medical opinion between the views of antagonistic parties, or carried with them the scientific world in a temporary enthusiasm of acceptance and admiration. They were often originated by men of learning and ability, and owed much of their success to the superior talent of their authors. They were usually ambitious in design, and seductive to the imagination. They claimed to open a wider field of knowledge, and to unlock the secrets of nature by some new formula that should dispense with the older and slower methods of plodding investigation. So long as they held out this prospect, they stimulated the hopes and attracted the interest of all. But after a time it appeared that the anticipations which they excited were not fulfilled, and that their promised results failed of realization. Then the theories themselves began to lose in importance; and, once started on the downward road, they receded more and more rapidly from view, until they finally passed out of sight and out of mind. They vanished into the limbo of departed spirits; and, if they ever show themselves at the present day, it is only to gratify a momentary curiosity as the relics of thought and opinion in former times. Like unsubstantial shadows, they flit across the pages of a forgotten literature, and the glimpses that we catch of them here and there are the only traces of their existence in the history of the past.

One of these phantoms, alive and flourishing a hundred years ago, was Buffon's theory of Organic Molecules.

Buffon was a remarkable instance of a man of splendid but temporary reputation. His ability commanded the attention of his contemporaries, and enlisted their interest and co-operation for natural history as no modern writer had done before. His immediate successors were fond of comparing him with Aristotle and Pliny; and, if we take their estimate of his reputation, there have been few men, either in science or literature, who ever acquired so great a popular renown. He was a member of the French

Academy, of the Academy of Sciences, of the Royal Societies of London and Edinburgh, and of the Academies of Berlin and Bologna. He was created a count by Louis XV., and was for forty years Superintendent of the Garden of Plants, which he more than doubled in extent, and where his statue was erected while he was yet alive. His great work on natural history was received as an authority from the date of its publication. Most of the European potentates sent him specimens for his museum or complimentary messages; strangers from abroad thought it a privilege to see him; and among the visitors at his country-house were Thomas Jefferson and the Prince of Prussia.

Notwithstanding this extraordinary success, Buffon's qualities could not secure a lasting reputation. He excited the enthusiasm of his readers by his eloquence and originality, but he did not give them much real instruction. This deficiency was felt even by his friends. Vicq d'Azyr, who pronounced his eulogy in the French Academy, and who calls him "one of the lights of the age, and an ornament to his country," says that he often "compels admiration, without convincing the judgment." According to Moreau, he "supplemented, by the brilliancy of his imagination, the imperfections of his knowledge," and showed himself "more remarkable for boldness of conception than for precision of ideas." There can be no doubt that he did much, in his own time, to popularize natural history among the educated and influential classes; but it was rather as an intellectual entertainment than a scientific pursuit, and his works are now nearly destitute of value for any purposes of reference or exact information.

It appears that Buffon's style as a writer was one of his principal charms for the readers of that day, and formed a large element in his success. It had a sort of poetic splendor, that would be thought quite inadmissible at present in a scientific work, and sometimes will hardly bear translation without a close approach to puerility; but at the period when he wrote, and in his hands, it magnified his subject and made an impression upon his readers. He often adopts this lofty tone at the opening of a chapter, and nearly always employs it for descriptive passages. He introduces the horse as "this proud and impetuous animal, the noblest conquest ever made by man, his companion in the fatigues of war and in the glory of combats." The elephant, he says, "makes the earth tremble beneath his footstep, uproots trees from the soil, and with one thrust of his body will batter down a wall." Buffon took great pains to cultivate and perfect this style. They say he would often repeat certain passages aloud, or get his friends to do so with himself for an audience, to enjoy the sound of his swelling sentences, and perhaps make some improvement in their construction. One of the extracts which he was especially fond of hearing in this way was his delineation of the feelings of the first man, when just awakened to consciousness and vitality. The newly created being is supposed to be telling his own story, as follows:[1] —

"I well remember that instant of mingled joy and anxiety, when I felt for the first time my extraordinary existence. I knew not what I was, where I was, nor whence I came. I opened my eyes. What an overflow of sensation! The daylight, the celestial vault, the green earth, the crystal transparency of the waters, all attracted me, excited me, and gave me an inexpressible sense of enjoyment." And so on, for seven or eight consecutive pages.

Another favorite passage was his picture of an Arabian

[1] Œuvres Complètes de Buffon. 2me édition. Paris, 1819. Tom. x. p. 358.

desert, introduced *apropos* of the natural history of the camel.[1]

"Imagine," he says, "a region without verdure or moisture, a blazing sun, the sky always arid, plains of sand, and mountains still more parched, where the eye searches in vain for any trace of a living creature; a land lifeless and, as it were, stripped by the winds, where nothing is to be seen but dead bones, scattered pebbles, and rocks standing or overturned; a barren desert, where the traveller never finds a shady resting-place, where he has no companionship, nor any thing to remind him of life in nature; absolute solitude, a thousand times more frightful than that of the forest, since even trees are living beings for the man who is there alone. In these void and measureless regions, more isolated, denuded, and desolate than any other, the empty space seems to enclose him like a tomb. The light of day, more gloomy than the shades of night, only serves to disclose his helplessness and destitution. It shows him the horror of his situation by enlarging before his eyes the boundaries of vacancy, and shuts him out from the habitable world by stretching around him an abyss of immensity which he can never hope to cross."

The satisfaction which Buffon felt in these descriptions does not seem to have been diminished in any way by the fact that he had never seen an Arabian desert, and that his story of the first man was entirely imaginary.

Buffon's "Natural History" was a work of very extensive design. It included an explanation of the structure and physical geography of the globe, the nature of man and animals, a general view of animals, vegetables, and minerals, the functions of nutrition and reproduction, an account of the varieties of the human race, and the different species

[1] Œuvres Complètes, tom. xv. p. 336.

of birds and quadrupeds, with a description of specimens in the Royal Museum.

He had a theory of the planetary system, which was one of his favorite doctrines, and which occupies the greater part of his first volume. He believed that the planets were originally a part of the solar sphere, from which they had been struck off at some time by the shock of a comet in collision with the sun.[1] The materials of different densities, propelled under this impulse with different velocities, he supposed would naturally collect into masses at corresponding distances from their starting-point, and so give rise to a series of planets, moving in successive orbits. Their axial rotation was accounted for by the obliquity with which the comet struck the sun, and the consequent twirling motion imparted to its detached portions.

Wild as this hypothesis sounds when stated by itself, it was advocated by its author with much force and ingenuity. Even mathematics came in for a share in its support, and it was fortified with arguments from the doctrine of probabilities. As all the planets move round the sun from west to east, and nearly in the same plane, they must all have been set in motion, according to Buffon, at the same time and by a single impulse. Otherwise, the chances would be as 64 to 1 that the six planets then known would not all move in the same direction, and as 7,962,624[2] to 1 that all their orbits would not be within seven and a half degrees of the same plane. He extended the calculation still farther, — to the size, distance, and density of the different planets, and to the relation in each between its density and the rapidity of its orbital motion. The figures did not all

---

[1] Œuvres Complètes, tom. i. p. 151.

[2] This was in 1750, when Saturn was supposed to be the outermost planet of the solar system. If the author had included in his calculations Uranus and Neptune, he might have made the chances in favor of his theory as 4,586,471,424 to 1.

come out exactly right; but in the case of Jupiter and Saturn the theoretical and actual proportions were very nearly alike, and the author observes with much satisfaction that "we cannot often obtain, in matters of pure conjecture, so close an approximation to reality as this."

But the most remarkable production of Buffon's genius was his system of organic molecules, with its associated theories of nutrition and reproduction. According to this view, the bodies of all living creatures, animal and vegetable, were composed of minute particles, differing in kind and configuration, but all possessing the general character of vital endowment. These were the organic molecules. As we call a compound group of hydrogen and oxygen atoms a molecule of water, or one of carbon, hydrogen, and oxygen a starch or glucose molecule, so Buffon seems to have imagined a still more complex molecule, having the properties of organic life. These molecules, when associated in the form of an organized body, gave to each part its specific character, and thus provided for the physiological activity of the whole. But they were themselves indestructible, and contained in their own substance the essential principle of vitality. When the animal or plant died and was decomposed, its organic molecules were at once ready to assume new combinations. When one animal fed upon another, or an animal upon a plant, the organic molecules of the creature used as food passed into the tissues of that which devoured it; and when an animal, in decomposing, enriched the soil and fed vegetation, its organic molecules were absorbed into the structure of the growing plant. Thus these molecules represented the living elements of the material world. Their quantity was never either increased or diminished; they only passed from one form of combination to another, always transferring their activity from the old organisms to the new. Consequently there was no such thing as destruction of life in nature, but only continual changes in its distribution. The organic molecules were an exhaustless reservoir of vitality, from which all living beings took their origin, and into which they all again returned.

But nutrition and generation require something more than the mere existence of organic molecules. For, even supposing these vitalized particles to be present in full abundance and variety, how could they be arranged in proper order within the plant or animal, so as to form its different organs or enable it to reproduce its like? This want was supplied by Buffon's celebrated idea of *interior moulds*. It is not easy to say exactly what he meant by this expression, except that it was something capable of arresting the molecules, and placing them in a certain combination. "In the same way," he says, "that we have moulds for giving to the outside of things any form we choose, suppose that nature can produce moulds which will determine, not only the external configuration, but also the internal structure of bodies formed in them."

Most readers will perhaps say that this definition of interior moulds is only an analogy, and does not include any intelligible idea of their construction. Buffon himself seems partly conscious of this, and makes a rather labored and metaphysical defence of the term, to show that there is nothing in it contradictory. However that may be, the "interior mould" is something which belongs to the entire body and also to each of its separate parts. The whole body is a mould which determines the position and arrangement of its different organs. Each organ and each part of an organ, when supplied with organic molecules from the food, selects those which are like its own, and rejects the rest; while it arranges the incoming molecules in proper order, and incorporates them with its substance, still preserving its original form and texture. In this way the body is nourished and grows, not by addition to its exterior, but by the intimate penetration and intus-susception of appropriate molecules throughout its mass; and all by means of its "interior moulds."

The mysterious phenomena of reproduction were fully explained by the aid of organic molecules and interior moulds. This explanation was even the simplest in cases which we usually consider the most wonderful; as where plants, worms, or polypes reproduce and multiply, without sexual generation, by division or budding. Suppose, in a polype, the organic molecules to be arranged throughout in the form of minute polypes, just as a cube of salt may be made up of an indefinite number of smaller cubes; then any part of such an organism, when separated from the rest, will reproduce an entire polype, because it already contains the whole of it in miniature. Something similar takes place in animals or plants which multiply by budding. In them the organic molecules absorbed with the food are distributed in due proportion until every part has received its full supply. After that, the surplus molecules are sent back from the different organs and collected in some particular locality, where they form a composite mass representing the materials of the whole body. Such a mass is a bud; and when separated from its stock, and placed where it can absorb nourishment for itself, it of course grows into an organism like its parent, because it is composed of the same kind of molecules, arranged in the same order and in the same proportions.

In sexual generation, like that of the higher animals, the process is more complicated. Here, again, every part of the body absorbs the organic molecules appropriate to itself, and moulds them into the substance of its own texture. But the excess of material, not required for the nourishment of the organs, and returned from them as superfluous, appears in the form of a liquid, and is deposited in a special reservoir. This liquid, made up of contributions from every part of the body, is the seminal fluid. It is a mingled organic extract of the entire frame, and thus contains all the materials necessary for reproduction. Its place of deposit is in the vesiculæ seminales.

The extreme activity of this seminal fluid, the concentrated essence of the animal organism, is proved by the formation of the spermatic animalcules. According to Buffon, these are not original or essential ingredients of the seminal fluid, but a product of its restless vitality, an incidental form in which its generative power is exhibited.[1] They represent the first assemblage of its organic molecules, as combined into moving particles large enough to be visible by the microscope.

But the female also produces a seminal fluid in the same manner and with the same general qualities as the male. Both fluids, male and female, contain all the elements common to either sex. The only difference between them is, that, each fluid being a special extract of the body from which it comes, that of the male contains molecules representing the male peculiarities, that of the female those representing the female peculiarities. Buffon entirely rejects the doctrine that viviparous animals produce by means of eggs. For him, the ovaries are testicles, the Graafian follicles are reservoirs of nutritious lymph, and the corpora lutea are glandular structures, which produce the female seminal fluid by a process of filtration.[2]

[1] Œuvres Complètes, tom. ix. p. 335. [2] Ibid., p. 345.

Thus Buffon accounts for reproduction in a very simple manner. Generation is only, in a higher degree, the act of nutrition, operating with the surplus of nutritive material. A mixture of the male and female seminal fluids contains all the organic elements of both sexes, and therefore represents the species. In the uterine cavity it finds its appropriate nidus and nourishment, and grows into an organism like that from which it came. The theory was supported by abundance of proof, for it explained to a charm most of the important phenomena of generation. 1st, During infancy and childhood there is no power of reproduction, because all the organic molecules are taken up by the growing tissues: it is only after puberty, when the body has acquired its full growth, that there is a surplus of organic material, to be filtered through the testicles in the form of seminal fluid. 2d, The young child or animal usually resembles both its parents, because its substance is derived from both the male and female seminal fluids. 3d, Creatures of small size, like insects, mice, and guinea-pigs, have a numerous progeny, because the demands of nutrition are easily satisfied, and there is a large superfluity left for reproduction. But those of greater bulk, such as man, the horse, the elephant, produce only a single young one at a time, and at long intervals, since the nourishment of their own bodies takes up most of the organic molecules of the food. All these facts were plainly in accordance with the new system.

Indeed, the only difficulty about Buffon's theory, if it is not a contradiction to say so, was that it made the matter too easy. If the seminal fluid of the male contains all the organic molecules necessary to make an embryo, why does it not make one? Why should not the male be self-producing? and why should not the female, on the other hand, have a progeny of little females, from her own seminal fluid, without requiring any thing further? According to the theory of organic molecules, every adult animal produces a fluid capable of generating young. But, in point of fact, the female must also receive the influence of the male; and, without the concourse of the other, each sex remains barren.

This is the great difficulty in Buffon's way; and, to do him justice, he does not try to conceal or evade it, but struggles with it manfully for sixteen octavo pages. The reader almost feels a sympathy for him, when his system of reproduction, after running so smoothly everywhere else, is arrested by so bulky an obstacle as the fact of sexuality. His explanation is full of ingenious hypotheses; but it is very complicated, and leaves in the chain of evidence so many gaps and contradictory possibilities, that it is hardly worth while to remember it. Nevertheless, the author thinks too well of the system to have his faith in it shaken by one or two imperfections. The main part of it, he says, he has "proved by so many facts and arguments, that to doubt it is hardly possible." He does not doubt it himself; and he avows that he "entertains no uncertainty whatever as to the basis of the theory, after careful scrutiny of its principles and the most rigorous estimate of its consequences and details."

Buffon's system has its widest extension and most splendid success in the dim regions of equivocal and spontaneous generation. His organic molecules are the genuine repositories of life, the interior moulds of particular organisms being only the means of arranging them in special forms. And when, by the death of an animal and the decomposition of its body, its organic molecules are released from the constraint of their interior moulds, they are set at liberty, living and active as ever, to be disseminated in any direction where chance may carry them. In this condition they are the universal seed of nature; without any character of specific organization, but containing all the prolific force of the bodies from which they came. After a time they are absorbed by some other mould, animal or vegetable, into which they transport their own indestructible qualities of nutrition and life. But during the interim, after leaving the old organism and before their incorporation into the new, they form by their re-union a multitude of living structures which do not belong to any of the ordinary or recognized species in nature. Such are the moving corpuscles of the seminal fluid, the microscopic fungi and infusorial animalcules of organic solutions, and, in their larger varieties, the eels of starch-paste and vinegar, and even earth-worms, mushrooms, and entozoa. All these creatures come by spontaneous generation. They are the connecting link in nature between the simple organic molecule on the one hand, and the complete animal or plant on the other. This exuberant force of production is always in activity. Notwithstanding the death of any number of existing creatures, the total quantity of life on the globe cannot be diminished even for a moment; for it continues uninterrupted in the organic molecules, which pass and repass from the animal to the vegetable, and from the vegetable to the animal, or produce in the interval an endless variety of spontaneous generations.

No doubt, Buffon was in love with his system, and admired it for its own sake. Still he had a fair share of what Condorcet calls the "sentiment of his own superiority," and ends one of his chapters in a strain that sounds, at this day, rather exalted. "I will add nothing more," he says, "to these reflections. To be accepted, or even to be comprehended, they require a profound acquaintance with nature, and complete freedom from prejudice; so that, for the majority of my readers, a longer explanation would be insufficient, and, for those who can understand me, it would be superfluous."

When Buffon's work appeared, it attracted at once the curiosity and applause of the public. The reviewers praised it as "an honor to both the age and the nation."[1] Hume said that its author "gave to things no human eye had seen a probability almost equivalent to proof;" and Needham declared that his investigations on the seminal fluid would be a subject of admiration "for centuries to come." All his readers, however, were not of that opinion. There were some who did not accept the doctrines of the book, and who feared that its popularity might have a bad effect. Buffon's high position, his magnificent style, and his confident tone of assertion, alarmed his adversaries, and made them dread his influence on the public mind. His theory, they said, was "filled with dazzling paradoxes, as disastrous to the principles of science as the mines of a skilful engineer to the solid ramparts of a fort." This difference led to a warm discussion. The Abbé de Lignac wrote an anonymous treatise against Buffon's work, under the title of "Letters to an American,"[2] where he criticised the system of moulds and molecules, and said, that, instead of describing natural objects, it contained only "the philosophical dreams of M. de Buffon." Buffon's friends, on the other hand, denounced the "American Letters" as a lampoon. They accused de Lignac of having written it

[1] Journal des Savans. Paris, Octobre, 1749.
[2] Lettres à un Ameriquain, sur l'Histoire Naturelle de M. de Buffon. Hambourg, 1751.

out of envy, and to punish Buffon for not showing more respect to his friend Réaumur. Deslandes called the book an imposture, because it was entitled " Letters to an American," when it had not been written, in the form of letters, to anybody; and because the title-page bore the imprint of Hambourg, though it was really printed at Paris. "That makes no difference," replied de Lignac: "the only title-page that can be called an imposture is one that promises something the book does not contain; like a Natural History, for instance, made up of disjointed and extravagant systems, without any thing natural about them."

Buffon's critics were especially dissatisfied with the organic molecules of the seminal fluid, and their wonderful capacity for producing an embryo. "If these molecules," they said, "are indifferently the elements of a plant, a man, a dog, or a polype, how can they combine to make either? We might as well go back to Epicurus and his atoms." One of them was asked, by an admirer of Buffon, which he would rather be, — the author of the "Histoire Naturelle," or that of the "Lettres Américaines." He replied that he would rather be an organic molecule, because then he would have more intelligence than either of them. Buffon held his own against these attacks without making any direct reply. His literary talent alone was almost enough to secure him the victory; and, if his opponents denied the truth of his theory, they generally had nothing better to offer in its place.

The only investigations at that time, which seriously threatened Buffon's system, were those of Haller on fecundation and development. They first appeared in 1753, four years after Buffon's first publication. Haller had examined the ovaries of sheep both before and after impregnation, and observed the condition of the Graafian follicles and corpora lutea. Since corpora lutea, according to Buffon, were glands for producing a seminal fluid, they must, of course, be formed before impregnation, and could only discharge their fluid at the time of conception. Haller declared that there were no corpora lutea in the ovaries during the period of heat, nor immediately after conception, at which time there was nothing to be seen but a simple orifice leading into a ruptured follicle. The glandular structure of the corpus luteum was only formed some days afterward, and therefore could not have furnished any thing essential to conception.[1] We now know the accuracy of these observations by Haller, and can hardly understand how those of Buffon should have been so preposterously incorrect. But at that time it was difficult to decide between them. Both were men of high scientific position, and the system of organic molecules was too promising and seductive to be surrendered at the first attack.

But as time went on, it began to suffer from its own defects. After thirty or forty years' trial, its promises had not produced any thing, its obscurities had not been cleared up, and the organic molecules were still as hypothetical as ever. One of Buffon's tenets for explaining reproduction by division was, that the simpler organisms, like plants and polypes, were composed of particles in the form of their own miniature; and he had argued at great length that this kind of structure, if we only thought so, was as simple as any other. Malesherbes[2] finally concluded that it would be of no use to think so, because the thing was impossible. "To be convinced of this," he says, "let any one try to

[1] Histoire de l'Académie Royale des Sciences. Année 1753, p. 134.
[2] Observations sur l'Histoire Naturelle de Buffon. Paris, 1798.

draw a picture of this pretended polype, entirely made up of similar polypes; or let him get a quantity of little wooden or ivory figures in the form of polypes, and then see whether he can put them together in any way so as to make an entire polype of the same shape. I defy him to do so."

Buffon's interior moulds had no better luck in the end. When it came down to genuine matter of fact, no amount of verbiage could make it plain what sort of things they were, nor how they operated. Malesherbes declared that the very idea of a mould is that of something which acts by its surfaces; and that consequently the hypothesis of interior moulds is a mere metaphysical subtlety, without any real application. "To make use of interior moulds, therefore, for explaining the process of nutrition or development, is to solve one difficulty by another: it accounts for one thing by the supposition of a second no more intelligible than the first. The truth is, it does not really give any explanation at all: it only answers, like the doctor in Molière, *Opium facit dormire quia habet facultatem dormitivam.*"

Another cause of decline for Buffon's system, toward the end of the last century, was the progress of opinion in regard to spontaneous generation. At that time there was much investigation on this subject, especially as to the infusorial animalcules. Needham and Spallanzani were the principal disputants in this field, and attracted to it, for some years, the attention of the scientific world; the result being, at last, a general conviction that Spallanzani's experiments had destroyed the hypothesis of spontaneous generation, at least for the infusoria. Buffon's system, therefore, diminished in importance, because this mysterious department of reproduction, which it had explained so well, no longer needed explanation. It thus lost one of its strongest claims to support; and, without being actually disproved, it was discredited to an extent which must have seriously impaired its scientific status.

Finally it received its death-blow, in 1827, from de Baer's discovery of the ovule in the ovarian follicle of mammalians.[1]

De Graaf had supposed, when he first described the ovarian follicles, a hundred and fifty years before, that they were really eggs; and this view was widely accepted. The ovaries of mammalians, like those of fishes and birds, were regarded as clusters of eggs held together by intervening tissue. But it was not easy to demonstrate, in man and quadrupeds, the discharge of these supposed eggs from the ovary. No observer had ever succeeded in finding one of them separated from its ovarian connections; and the earliest ovum discoverable in the uterus, or Fallopian tube, was, of course, considerably smaller than the ripe follicle in the ovary. This threw a doubt over the doctrine of de Graaf, and left room for Buffon's assertion that there were no such things as eggs in the mammalian female; that her so-called ovaries were testicles, as much so as those of the male; and that she produced a seminal fluid composed of organic molecules similar to his.

This view was so interwoven with Buffon's whole system, that one could hardly exist without the other. And when the microscopic ovum and its mode of exit from the ovary were at last fully brought to light, so that any one could see them when he chose, the discussion necessarily came to an end. The female testicles and seminal fluid, with their glandular corpora lutea, no longer had a place in court: every thing connected with them shared in the collapse; and the entire company of organic molecules, repro-

[1] K. E. von Baer: De Ovi mammalium et hominis genesi. Lipsiæ, 1827.

ductive extracts, composite organisms, and interior moulds, disappeared together, like actors at the end of a play.

Another set of ideas, once in high repute but now long obsolete, are those of Bonnet on the "Inclusion of Germs."
Bonnet was of a wealthy and influential family in Geneva, where he passed the whole of his life, engaged in the study of natural history and kindred subjects. He became widely known for his attainments in this direction, and was connected with nearly all the learned societies of Europe. He was the friend and correspondent of Réaumur, Haller, Spallanzani, Trembley, and de Saussure. Spallanzani communicated to him in manuscript, for perusal and advice, his own researches on the generation of infusoria; and, when the volume appeared, he printed, as part of it, Bonnet's letters in reply.[1]

The personal character of Bonnet had traits of superior excellence, and secured him a large share of attachment and esteem. His disinterestedness and modesty, in regard to his own opinions, were well known, and were often the subject of eulogy. His own discoveries in entomology were of acknowledged merit; but he attributed them mainly to the teaching of Réaumur, because he had been inspired by the example and encouragement of this master, under whom he felt "proud to enroll himself as a pupil." Spallanzani calls him the "Philosopher of Geneva" and the "sublime author of the 'Contemplation of Nature.'" His first publications appeared in 1745, and were followed by others at various dates until 1770, after which they were collected in a series of eighteen volumes.[2]

Bonnet presented the curious spectacle of a man under

[1] Spallanzani: Opuscoli di fisica animale et vegetabile.
[2] Ch. Bonnet: Œuvres d'Histoire Naturelle et de Philosophie. Neuchatel, 1779.

the varying influence of two opposite tendencies, — at one time an exact and painstaking observer of nature; at another, an irresponsible wanderer in the vaguest of speculations. His earliest scientific work was an experimental investigation on the reproduction of *aphides*, by which he first demonstrated with certainty the non-sexual viviparous propagation of these insects.[1] It is hardly possible to imagine a more thorough, direct, and conscientious method than that adopted by Bonnet in these observations. He took the young aphis just expelled from the body of its parent, placed it upon a sprig of fresh leaves stuck in a pot of earth and covered with a bell-glass, and then watched his solitary prisoner for twenty-one days, until it had grown to maturity, and had produced, without fecundation, ninety-five young ones. During this time he kept exact records of the insect's moulting, and also of the new births, most of which took place under his own eye. He repeated the trial with aphides of different species, and with those which had themselves been produced in captivity; and he succeeded in raising them in this way, without fecundation, for eight successive generations. For these observations he received the honor of an election as Correspondent of the French Academy of Sciences. He also made valuable contributions to natural history on the multiplication of aquatic worms by division, on the growth of plants, on the function of leaves, and on the habits of insects.

On the other hand, he constructed, out of his own meditations, a graduated system of things visible and invisible, which he called the " Scale of Being," and which formed the subject of one of his later works, published under the title "Contemplation de la Nature." It embraced the structure of the entire universe, from the simple elements

[1] Œuvres de Bonnet, tom. i. p. 19.

of inorganic matter to the "celestial hierarchies" of other worlds. According to this system, there are no gaps or sudden variations in nature, but only a continuous series of combinations and modifications in ascending grade. Every thing is immediately connected with that which precedes and with that which is to follow. At the beginning of the volume there is inserted, for the reader's convenience, a printed table exhibiting the regular succession of natural objects, from gaseous substances up to man. The inorganic elements, in various grades of combination, lead to the composite earths, sulphurs, and metallic bodies; thence, through the vitriols or metallic salts, to ordinary saline substances in crystalline form; and thence to compound rocks of heterogeneous structure, like granite and sandstones. Immediately above these are rocks, like slate, talc, and asbestos, which exhibit a laminated or fibrous texture, and which may therefore be considered as "the connecting link between inorganic and organic solid bodies."

This point is especially important in Bonnet's scheme, for it marks the passage from dead matter to living organisms. It is not altogether what its author could wish; but he makes the best of it by quietly putting the burden of proof on posterity. "It must be acknowledged," he says, "that the transition here is not quite so perfect as in many other cases. It looks as though nature in this instance had made a leap. But the gap will undoubtedly be filled up as our knowledge increases in precision and extent."

Once fairly started among organized structures, the scale rises easily through corals, lichens, and mushrooms, to plants in general; through the sensitive plant to animals; from insects to snails, serpents, eels, and fish; through flying-fish to birds; from birds to bats and quadrupeds; and so on, to monkeys, apes, and man. But this extended gradation is not enough for Bonnet. He has also the "Celestial Hierarchies" and the "Gradation of the Worlds;" some of which, he believes, are as superior in physical and moral organization to our earth as man is above the ape.

The frontispiece to Bonnet's work is an engraved portrait of himself, taken, as he tells us, when he was "plunged in deep meditation on the restitution and perfectibility of living creatures." It renders, he thinks, his "reflective expression" with much success; and it is certainly fortunate that the artist saw him just when he happened to be meditating on so abstruse a subject:

Bonnet's theory of reproduction, or the "Inclusion of Germs," appears mainly in his work entitled "Considérations sur les Corps Organisés," with some additional amplifications in the "Contemplation de la Nature." His thoughts were turned in this direction partly by Trembley's recent discovery of the multiplication of fresh-water polypes by division, and partly by his own observations on the reproduction of aquatic worms and aphides. He was further strengthened in his opinion by Haller's observations, which appeared in 1758, on the formation of the chick in the fowl's egg.

There were, about that time, three different views entertained as to the nature of sexual generation, —

First, The spermatic animalcules, one or more in number, were arrested in the uterus, and there became developed into embryos. According to this view, the male parent furnished the germ, the female only providing for its care and nourishment.

Second, The seminal fluids of both sexes, when mingled together, produced the embryo, *de novo*, by a mutual arrangement of their organic particles. This was the doctrine of Buffon, and attributed an equal and similar action to both parents.

Third, The germ pre-existed in the ovary and egg of the female, and was only stimulated to further development by the seminal fluid of the male.

Bonnet accepted the third of these doctrines. He believed in the pre-existence of germs, but he gave to the idea a much wider extent and significance. The germ, as he understood it, was not only a structure capable of being developed into a living creature, but actually contained, within itself and ready formed, all the parts of the future organism. These parts were exceedingly minute, delicate, and transparent; but they nevertheless existed, though as yet imperceptible to the senses. It was only their small size, their softness and transparency, which prevented their being seen. After impregnation they grew larger, more consistent, and more opaque, and then first became recognizable as the organs of the embryo. These organs, accordingly, were not things of new formation. They only showed plainly, at a certain period of development, what had previously existed in invisible form. The author extended this doctrine without hesitation to its farthest limits. For him, every seed contains a plant in miniature. In an animal germ, every organ and tissue, without exception, is already present,—the heart, the vessels, the liver, the kidneys, the intestine, the nerves, and the brain. Consequently there is no such thing as generation, if by that we understand the production of a new being, or even of a new organ. What appears to be such is only the development of something that was in existence beforehand.

These views were adopted by some, and rejected by others; the physiological theorizers of the day dividing into two parties, known as the *epigenesists* and the *evolutionists*. The epigenesists were those who believed, like Buffon, that the embryo was a new formation, produced by the union of molecules and the successive organization of different parts. The evolutionists,[1] on the other hand, maintained, with Bonnet, the pre-existence of germs, and insisted that nothing in the embryo was newly formed, but only enlarged and solidified by development.

These notions were sustained in some degree by the discoveries of Haller in the incubation of the fowl's egg. Haller had studied the development of the chick, day by day, and had noted the appearance of its different organs and envelopes. His masterly observations on the yolk-sac, and its connection with the embryo, were claimed by the evolutionists to be decisive as to the pre-existence of the germ and its essential parts. "The membrane enclosing the yolk," says Bonnet, "is continuous with the intestine of the chick. The arteries and veins of the yolk are branches of the mesenteric vessels of the fœtus. The blood circulating in the yolk-vessels receives its impulse from the fœtal heart. The yolk is, therefore, essentially a part of the chick. But the yolk exists in the egg before impregnation: consequently the chick also exists in the egg before impregnation. . . . This proves incontestably that the embryo is at first a part of the parent, and that its existence is anterior to the time of conception."

Bonnet's assumption of the pre-existence of the embryo, as well as of the germ, was not without a certain plausibility. The organs of the embryo, when first formed, are, in fact, not only small, but also colorless and transparent; so that they are seen with considerable difficulty, and may even be overlooked for some time after they are really present. Bonnet says this is the mistake made by the epigenesists. "They estimate the time when the parts of an animal are

---

[1] It is hardly necessary to say that this term, as used during the last century, had a meaning different from that which attaches to it now.

first created by that at which they first become visible; and whatever cannot be seen they assume has no existence." But the organs in an imperfect embryo might very easily have been there, though not sufficiently distinct to meet the eye. "Do you wish," he says, "a short and ready demonstration of this? When the lung of an embryo chick is first perceptible to the senses, it already measures ten one-hundredths of an inch. It is evident that it would have been visible when only half that size, were it not for the extreme transparency of its tissue."

And so of all the other parts. Bonnet has more faith than the epigenesists. He believes that the organs of the embryo were there before he could see them, and that they even existed from the beginning, though invisible, in the unimpregnated egg.

But if the germ contains all the organs of the chick, ready formed, only small and undeveloped, it contains ovaries as well as the rest; and these ovaries contain the germs of the next generation. It is plain the series cannot end here. The germs of the second generation are chicks in miniature, as much as those of the first. Consequently they contain their own ovaries and germs, which, in turn, contain others still; and so on, without any definable limit, one set of germs within the other in geometrical progression, each germ containing, not only its own embryo, but also all the embryos that can possibly be produced from it in unnumbered generations hereafter.

Bonnet does not shrink from the magnitude of this spectacle. He rather glories in it, and enjoys the contemplation of its unlimited perspective. "For my part," he says, "I like to extend as far as possible the limits of creation. There is a pleasure in the thought of this magnificent series of organized beings, enclosed, like so many little worlds, one within the other. I see them receding by degrees, diminishing in regular order, and losing themselves at last in an impenetrable night. It gives me a sense of satisfaction, when I look at an acorn, to see in it the germ of a majestic oak that, centuries hence, will shelter the birds and the beasts under its branches; or, still more, to behold in Emilia's bosom the germ of the hero, who, thousands of years hereafter, will found an empire; or perhaps of the philosopher who shall discover to mankind the cause of gravitation, the mystery of reproduction, or the mechanism of our existence." [1]

This prospect, however, for the germs of the future, which Bonnet found so attractive, was not as satisfactory to some of his contemporaries. They even haggled with him about the past. If the age of the world, as generally admitted, was six thousand years, and if each generation of man is completed in thirty years' time, that would give two hundred successive generations since Adam and Eve, with about a thousand million human beings at present on the earth. All these inconceivable numbers must have been contained, under the form of germs, in the ovaries of Eve. The minuteness of the embryos of the two-hundredth generation, when enclosed in the body of our first parent, was declared to be something incredible. Still more so for plants which grow and fructify every year. Hartsœker calculated that the relation in size between the first wheat-grain, and that to be produced at the end of six thousand years, would be expressed by the figure of unity, *followed by thirty thousand zeros*.

All this did not give Bonnet or his friends any discomfort. Such terrific estimates may serve to confound the imagination; but logic finds a sure refuge in the infinite or

[1] Œuvres de Bonnet, tom. v. p. 206.

indefinite divisibility of matter. Hartsœker and his followers, according to Bonnet, put the senses and the imagination in place of the understanding. They expect to see and touch, so to speak, what can only be comprehended by an effort of the mind. The hypothesis of the inclusion of germs is above the level of such calculations, and is, he concludes, a "splendid triumph of pure reason over the senses."

Bonnet's theory is by no means confined to the simple hypothesis of one germ included within another. It is an extensive and elaborate system, ramifying into all departments of natural history and physiology. The included germs not only exist before fecundation, but they are nourished and grow, in a way analogous to that of adult bodies. There can be no doubt of this; because the fowl's eggs certainly increase in volume before fecundation, and are to be seen in the ovary, of many different sizes and grades. But the yolk is essentially a part of the chick: therefore, if the yolk grows before fecundation, the embryo does the same. Besides, in the theory of inclusion, the contained germs diminish progressively, one within the other, to an inconceivable minuteness. The outermost germ is, of course, the largest, and is nearly ready for impregnation. But when it has been fecundated, and has expanded into the body of an adult female, the next included set of germs have, in turn, come to be as large as the other was before; and so on, in continuous order, the smaller germs gradually moving up in the scale of magnitude, to be ready for impregnation when their time arrives. The process of nutrition and growth must therefore be going on in these embryonic organisms, and has been going on in them ever since the creation.

But how is the nutrition of the germs provided for, and what are the materials used in their growth?

It is well known, says Bonnet, that the different parts of the body are not nourished directly by the blood. A more delicate and colorless fluid, called "lymph," separated from the blood by appropriate organs, serves for the immediate supply of the tissues. And as these tissues vary in fineness and consistency, we can hardly doubt that the secretory organs prepare for them different kinds of lymph, more or less attenuated to correspond with their delicacy of structure. Some tissues, as, for example, the brain and nerves, must require a lymph many times finer and more subtle than that employed for the grosser parts.

But the tissues of the included germs are finer still. If we admit that there is no such thing as real generation, and that every thing has been preformed from the beginning, then the germs which are to make their appearance a thousand years hence have now actually within them, and incalculably minute, all the parts characteristic of the species. Imagine the heart, the brain, the nerves, and the stomach of such an embryo. Certainly, no lymph could penetrate their tissues, or be of the least use for their nutrition. What fluid is there in the body sufficiently attenuated and ethereal to insinuate its particles into a structure of such delicacy, and incorporate them with its substance? There is only one, and Bonnet secures it. This precious and double-distilled material, which is alone capable of nourishing an unfecundated germ, is the NERVOUS FLUID.

The nervous fluid was not an invention of Bonnet's. It was generally acknowledged, in his time, as the immediate cause of nervous action. Like the electric fluid, it was intangible and invisible, and was known only by its effects. Its invisibility was due to its vaporous or aerial constitution; and this enabled it to circulate in the delicate passages of the nervous tissue, which were themselves invisible, owing to their minuteness. The nervous fluid, or "animal spirits" so called, after being elaborated in the brain, was disseminated through the nervous channels to the various organs of sense and motion. Its existence was proved by a very simple fact. If you tie a nerve going to a muscle, the muscle is paralyzed. How could a ligature cut off the nervous influence, except by arresting the passage of a fluid? There was, therefore, little or no doubt in regard to it.

The nervous fluid, in Bonnet's system, was not entirely dissipated in the exercise of sensibility and motion. The unconsumed portion, after returning to the brain, was again sent out by similar channels to be used in nutrition. It was carried by the maternal nerves to the ovaries, and there distributed to the germs "of the first order," or those nearest maturity.[1] Elaborated anew, and still further spiritualized by the organs of this delicate structure, it supplied a finer nourishment to the germs of the second order: here it underwent a similar preparation for those of the third order; and so on, to the belated and unfortunate germ which "would not be ready for fecundation till the end of the world."

This staggering array of infinitesimals, in structure and function, only serves to increase the author's admiration, and strengthen his confidence. "What an abyss of wonders," he says, "in the human brain! Even a single fibre of the brain is itself an abyss. What shall we say, then, if all these mysteries, existing in the adult body, are repeated and concentrated in so many organized atoms, enclosed one within the other, and growing constantly smaller in indefinite progression?"

The act of impregnation, in reproduction by sexes, comes

[1] Œuvres de Bonnet, tom. x. p. 10.

in for a very easy and natural explanation. The nervous fluid is the appropriate nourishment for an unfecundated germ. When this germ becomes a fœtus it will be supplied with lymph, and its organs will grow like those of the adult. But how can it make the transition from one state to the other, and how are the finely woven tissues of the germ-embryo to be made capable of admitting the grosser substance of the lymph? This is exactly the function of the seminal fluid. It is not only a source of nourishment for the mature germ, but it has a special stimulating influence on the embryonic heart. The pulsation of the heart in the germ-embryo is very faint, and only serves for the circulation of the nervous fluid. Consequently the unfecundated germ increases but slowly, and never gets beyond a certain stage of development. But when the seminal fluid comes in contact with it, and penetrates its substance, the heart's pulsation is increased in force. This causes an expansion of the circulatory channels, an increased volume of the circulating fluid, and consequently more rapid nourishment; and thus the limit is passed between the condition of the germ-embryo, whose tissues only admit nervous fluid, and that of the growing fœtus, that can thrive upon the lymph. Once begun, the process naturally continues until the body has attained its full development.

Still the theory of inclusion was not altogether free from difficulty. One of its most troublesome problems was that of intestinal worms and other parasites. Bonnet wrote a special treatise on the *tænia*, partly devoted to different theories of its origin. He rejects, of course, the idea of equivocal or spontaneous generation. But if, like every thing else, it comes from pre-existing germs, are these germs introduced into the intestine from without, or did they exist beforehand in the embryonic human body itself? If so, the germ of the first tape-worm, and, in fact, of all which have existed since, must have been included in the body of the first man. There are objections to be met in either case, but the most singular are those connected with the second.

In this view, the tænia has a contemporaneous origin with man; that is, it already existed in Adam, and so has passed into his posterity. But, according to Genesis, the animals were all created before Adam: therefore the tænia must have first lived in a foreign element; but the fact is, we never find it except in the intestines. Besides, how can we suppose such a pest to have been placed in the body of our innocent progenitor, living in the earthly paradise of the Garden of Eden? Valisnieri, who adopted the contemporaneous theory, had an odd answer to this objection. He assumed, that, before Adam's fall, the intestinal worms did not do him any harm, but, on the contrary, were rather useful, in "consuming the superfluous humors and gently stimulating the intestinal walls." Bonnet thinks this explanation a trifle far-fetched, and doubts whether Adam, in his state of innocence, had any superfluous humors, or needed any tape-worm to excite his intestines. He prefers the idea of Leclerc, that the intestinal worms, which are now so troublesome, only existed at first in Adam under the form of eggs, and were not hatched until after his disobedience.[1]

But the line of descent for tape-worms is not yet quite clear. Granting that they existed in Adam, either as eggs or worms, how did they get into the body of Eve, through which they must have passed for transmission to posterity? The most direct route from the intestine, according to Valisnieri, would be through the thoracic duct, which

[1] Œuvres de Bonnet, tom. iii. p. 140.

"mounts upwards along the ribs" to that which was taken for the body of Eve. Bonnet is inclined to adopt a simpler and less miraculous method of transfer, by supposing that the eggs of the tænia are so small that they can pass through the blood-vessels, and thus reach the vesiculæ seminales. "In this way," he says, with a puzzling kind of *naïveté*, "we can explain the whole thing without much difficulty, and *without having recourse to divine power.*"

However, the author regarded the introduction of parasitic germs from without as, on the whole, the more probable theory of their origin, especially as the weight of evidence was already in that direction.

The system of Bonnet has applications to many different points of general physiology; such as the resemblance of the young animal to both parents, the production of hybrids and monstrosities, and in particular to non-sexual generation, or multiplication by budding and division; all of which it explained with more or less ingenuity and success. It had an apparent support in several recent and important observations; such as the existence in certain seeds of an embryo plant, the multiplication by interior germs of *Volvox globator*, the reproduction of *aphides* without impregnation, and Haller's researches on the embryo chick. Haller himself defended it from a variety of objections, and adduced a number of arguments in its favor.[1] But the theory itself went so far beyond the limits of possible demonstration, that its basis of reality bore no calculable proportion to its towering superstructure. It discounted the results of embryological discovery at too exorbitant a rate; and it could not afford to have any of them turn out differently from what was expected. This happened in the case of the mammalian egg. Subsequent observations, made

---

[1] Elementa Physiologiæ Corporis Humani, 1766. Tom. viii. sect. ii. art. xxix.

with better facilities, showed that the organs of the fœtus do not pre-exist in the ovum in the sense imagined by Bonnet. The impregnated ovum has, it is true, a remarkable property, by which each part subsequently assumes a definite shape, and becomes a distinct organ. But the actual configuration of these organs is certainly not present in the ovum, because we can now see its original structure, as well as the intermediate forms which afterward give place to those of the embryo. The system of the inclusion of germs, in the time of its glory, embraced the whole history of the world from beginning to end, and the mechanism of growth and reproduction throughout organic nature; and enlisted in its behalf the discoveries of scientific research, as well as the results of logical deduction. But it came to grief when the study of embryology had reached a certain stage in its development, by demonstrating the actual constitution of the mammalian egg.

LECTURE III.

NERVOUS DEGENERATIONS AND THE THEORY OF SIR CHARLES BELL.

Mr. President and Gentlemen, — It is now a little over forty years since the beginning of an investigation which has produced important changes in the study of the nervous system. It relates to the alteration of structure taking place in nerves after they have been divided by transverse section.

The manner in which attention was first directed to this point is as follows: It had long been known, that, when a nerve was cut at any point, the immediate effect was a suspension of its functional activity. If it were a motor nerve which had been divided, voluntary motion was abolished in the parts below; and, if a sensitive nerve, there was loss of sensibility in the region where it was distributed. But this interruption of nervous activity was not always permanent. In some cases the power of motion or sensation would return in the affected parts; and the ends of the nerve were then found to be re-united by cicatricial tissue, in which there were connecting nerve-fibres of new formation. This re-union of a divided nerve, and the mode in which its structure and functions were re-established, formed the subject of much laborious investigation.

The observations in question were mainly directed to

the cicatrix between the ends of the divided nerve. This was naturally supposed to contain the key of the whole secret, and consequently received most of the attention given to the subject. But in 1839 Nasse [1] extended his examination to the nerve-fibres at a lower level. He divided the sciatic nerve in frogs and rabbits, and, after waiting from two to five months, found that the fibres in the separated portion of the nerve had become altered in structure. They were of irregular form, granular, and semi-opaque; they often contained what seemed to be minute fat-globules, from the degeneration of their myeline; and they appeared to undergo, in this way, a gradual atrophy. Nasse's observations were almost immediately followed by those of Gunther and Schön,[2] which were to the same effect. They were made on the divided sciatic nerve in rabbits, from twelve hours to one year after the operation. They showed that a nerve, after some days' separation from its nervous centre, loses its irritability, so that its galvanization will no longer produce contraction in the muscles below, and that this loss of irritability coincides in time with the granular degeneration of its fibres.

All these researches were connected with the question of the re-establishment of the nervous continuity, and they did not extend for any great distance below the point of section. Thus far, whenever observers use the term "regeneration," as applied to divided nerves or nerve-fibres, they mean the formation of new connections in the cicatricial tissue. It was supposed, that, when this re-union took place, the fibres below were restored to their normal condition, the nerve being enabled, in this way, to resume

---

[1] Archiv für Anatomie, Physiologie und wissenschaftliche Medicin. Berlin, 1839, p. 405.
[2] Ibid., 1840, p. 270.

its functions. The fact of degeneration was considered as of secondary consequence, as limited in extent, and often of temporary duration.

But ten years later the subject was taken up afresh by Augustus Waller, and was treated by him in a manner so original and so thorough, that it at once assumed a character of the first importance. Waller began his professional life in England as a practitioner of medicine; but he had a strong taste for independent research, and, after being for a time professor of physiology in the medical school at Birmingham, he transferred his residence to the Continent, where he remained for most of his subsequent life, and where the greater part of his scientific work was accomplished. For the anatomical study of muscular and nervous fibres he employed the tongue of the living frog. This gave him a great advantage; since the extensibility and transparency of the organ enabled him to examine its internal parts in their normal place and condition, undisturbed by the contact of artificial reagents. He applied this method to the observation of nerve-fibres and the change which they undergo after division; and thus followed their granular degeneration in the living animal, from day to day, under the microscope.

One of his earliest and most important results was that the degeneration, in the fibres of a divided nerve below the point of section, extends through its whole length, reaching to its ultimate ramifications. After division of the glosso-pharyngeal nerve in the upper part of the neck, he saw its degenerated fibres in the tongue, and even in the papillæ of the mucous membrane. He also established the fact, that, in a nerve once divided, the degeneration of the affected fibres is final, and that, when the nervous functions are restored, it is by a process of regeneration;

that is, the growth of new fibres throughout the separated portion of the nerve.

But the principal merit in Waller's discovery was something additional. He did not rest content with its value as an acquisition: he made it serve as an instrument. He saw in the degeneration of divided nerve-fibres a means of investigation, which he at once put to further use. A nerve is often made up of fibres from different sources. It receives branches of communication from other nerves; and, in the inosculating filaments between them, there may be fibres going or coming in either direction. In the nervous trunks and branches these fibres are for the most part hopelessly intermingled; and it is impossible to follow them for any distance by ordinary dissection. But by dividing, in the living animal, a nervous trunk or branch near its origin, Waller could afterward trace its degenerated fibres, however erratic their course, or whatever their distribution. He communicated his results to the French Academy of Sciences,[1] under the title "*A new Method for the Study of the Nervous System, adapted for investigating the Anatomical Distribution of the Nerves.*"

The first application of the new method was in the frog's tongue. This organ is supplied by two nerves, corresponding respectively to the glosso-pharyngeal and the hypoglossal. After dividing one of these nerves, and allowing time for its subsequent degeneration, Waller could determine the course and destination of all its fibres. He followed them throughout the frequent anastomoses of the two nerves, where the altered fibres could always be distinguished from the normal ones around them; and the two kinds were so different in appearance, that, as he expressed it, "there could be no doubt whatever as to the origin of either."

---

[1] Comptes rendus de l'Académie des Sciences. Paris, 1851 Tome xxxiii. p. 606.

These results were obtained by repeated and accurate experimental investigation; and they have been fully corroborated by subsequent observers down to Vulpian,[1] in 1866, and Ranvier,[2] in 1878. According to Vulpian, six weeks after the division of the sciatic nerve in a dog, no unaltered nerve-fibres could be found in the muscles of the corresponding foot; and, in similar experiments by Ranvier on the rabbit, the terminal ramifications of the nerve in the muscles of the leg contained only nerve-fibres with disorganized and granular myeline.

The advantages in Waller's mode of procedure were so obvious and so readily verified, that it was generally adopted, and received the name of the "Wallerian method." It at once yielded in the hands of its author much interesting information. After division of the pneumogastric nerve on one side, in cats, at the level of the larynx, granular degeneration was found throughout its peripheral fibres, except in those joining it below from the inferior cervical ganglion of the sympathetic, which remained normal. Above this junction the nerve was wholly disorganized; below, it contained a mixture of fibres, degenerated and normal. It was thus made evident that the inosculating branches between the two nerves are given by the sympathetic to the pneumogastric, and not by the pneumogastric to the sympathetic.

Physiologists had often been at variance as to the origin and character of the *chorda tympani*,—that remarkable filament of communication between the facial and lingual nerves. Was it a branch of the facial nerve going to the lingual, or was it an offshoot from the lingual returning to join the facial? This point was determined by the method

---

[1] Leçons sur la Physiologie du Système Nerveux. Paris, 1866, p. 243.
[2] Leçons sur l'Histologie du Système Nerveux. Paris, 1878. Tome ii. p. 349.

of section. Waller introduced his stilet into the middle ear, and divided the filament where it crosses the membrana tympani. Fifteen or twenty days afterward, nearly all its fibres were degenerated between the point of section and the lingual nerve; that is, in the direction of their peripheral distribution. The chorda tympani was consequently shown to be a branch of communication *from* the facial nerve *to* the lingual.

In this way many doubtful questions of nervous distribution received their solution. The granular degeneration of the divided nerve-fibres became an artificial aid to the anatomist, of the same value as the injection of minute blood-vessels with opaque pigment. Even a single disorganized fibre could be distinguished among a crowd of normal ones; and, in a mixture of the two, their numerical proportion could be fairly determined.

But Waller did not remain satisfied with this success. He determined to apply the same method to the study of the nerve-roots;[1] and in doing so he opened a new chapter of nervous physiology, and greatly enlarged the field of his operations.

The history of these experiments is so well known, that they only need to be stated in a few words. Waller had already shown, that, if a spinal nerve be divided at its exit from the vertebral canal, all its peripheral parts suffer degeneration. Motor and sensitive fibres are alike involved, and can be traced, in their disorganized condition, to their ultimate termination in the muscles and the integument. On continuing the same mode of investigation within the vertebral canal, if the anterior root only were divided, the

---

[1] His second communication to the French Academy was entitled "Experimental Researches on the structure and functions of the ganglia." Comptes rendus de l'Académie des Sciences. Paris, 1852. Tome xxxiv. p. 524.

motor fibres alone degenerated in the trunk and branches of the nerve, while its sensitive fibres remained normal; if the posterior root were divided outside its ganglion, the sensitive fibres of the nerve degenerated, and the motor fibres were unaffected; and in both cases the portion of the nerve-root still connected with the spinal cord preserved its normal structure. But, if the posterior root were divided above the situation of its ganglion, two remarkable consequences followed,— first, there was no degeneration of the nerve outside the ganglion; and, secondly, that portion of the nerve-root separated from the ganglion, but still united with the spinal cord, suffered complete degeneration.

These effects were repeatedly observed, under varied conditions, until there was no doubt in regard to them. In the frog, when both roots of a sciatic nerve were divided above the level of the ganglion, there was, of course, complete loss of both motion and sensibility in the paralyzed limb; but, while all its muscular nerve-fibres degenerated, the sensitive fibres distributed to the skin exhibited no alteration, even at the end of two months. Similar experiments on warm-blooded animals turned out in the same way. In dogs and cats, Waller utilized for this purpose the exceptional position of the second cervical nerve, the two roots of which remain distinct in these animals until after their emergence from the intervertebral foramina. He could thus operate on the nerve-roots without opening the vertebral canal; and by this means he verified Magendie's discovery, that section of the anterior root destroys the power of motion, and that of the posterior root the power of sensibility. But he demonstrated, furthermore, that, when both roots are severed above the level of the ganglion, all the motor fibres of the nerve degenerate, and all its sensitive fibres remain sound.

It is therefore evident that the immediate loss of action in a divided nerve, and its structural degeneration, are two independent phenomena, having nothing to do with each other. A motor nerve, when cut, ceases to excite voluntary motion, not because its fibres are incapacitated by degeneration, but because their physical continuity is broken at one point; and a sensitive nerve, under like conditions, can no longer communicate with the sensorium, for the same reason. The proof of this is, that, if we galvanize the motor nerve below the point of section, it at once calls out the action of its muscles; and a like stimulus, applied to the sensitive nerve above its point of section, will be transmitted to the brain, and produce the effect of a sensation. This is the state of things immediately after the nerve has been cut.

But at the end of some days a different condition shows itself in the peripheral part of the divided nerve. Its fibres then become unable to re-act under artificial stimulus, and their galvanization no longer produces muscular contraction. This is due to their progressive degeneration, which requires time for its development, but which, at a certain stage, destroys the functional activity of the nerve-fibres. Simple paralysis of motion and sensation is consequently the *immediate* result of division of a spinal nerve. Its degeneration, with loss of irritability, is a *subsequent* result. The independence of the two is especially manifest in Waller's experiments on the spinal nerve roots. When both are divided above the level of the ganglion, the motor fibres of the anterior root become degenerated throughout their distribution in the paralyzed parts; but the sensitive fibres from the posterior root remain everywhere sound in the integument of the same region.

From these facts Waller concluded that the two sets of

nerve-fibres had different centres of nutrition; that the motor fibres of the anterior root were dependent, for structural integrity, on their connection with the spinal cord, while the sensitive fibres of the posterior root depended on their connection with the spinal ganglion. He was fully justified in claiming that his results were unequivocal, and "calculated to throw a new light on the relations of the nervous system."

For Waller's communications on this subject to the French Academy, he was awarded, in 1856, the Montyon Prize of Experimental Physiology, by a committee consisting of Flourens, Rayer, Serres, Milne-Edwards, and Bernard. In reporting the decision of the committee, the chairman, Bernard, begins with the following words: "The discoveries usually presented to us are of two different kinds. In some, the facts have been already, to a certain extent, anticipated, and are mainly serviceable in their new form by increasing the precision of our knowledge, and by elucidating subjects previously obscure. Those of the other kind, on the contrary, are unexpected. The principal feature in them is their novelty; and they enlarge the boundaries of science, not merely by the solution of particular questions, but still more by the ideas and suggestions which they originate. It is to the last of these two categories that the present essay belongs."

It soon appeared that the committee were not mistaken in their estimate. The immediate result of Waller's researches was the discovery of the so-called "centres of nutrition" for nerve-fibres of different kinds. The existence of these centres of nutrition was fully demonstrated; although it was not quite plain, and has not even yet become so, in what way their influence is exerted. But there was also the additional discovery that the degeneration of divided nerve-fibres may take place in opposite directions. It may be *centrifugal*, that is, from the point of section to the periphery; or it may be *centripetal*, that is, from the point of section toward the spinal cord and brain. This fact afterward became the most prolific source of further investigation.

Up to this point Waller's experiments had been confined to the spinal nerves and nerve-roots. In his last communication to the French Academy,[1] he goes a step farther, and traces the course of similar degenerations in the cord itself. He divided the spinal cord, in a dog, between the third and fourth lumbar vertebræ. After twenty days the cord was found to be re-united at the line of section. But above the section the posterior columns were disorganized for the space of about two vertebræ; while in the inferior segment the same columns, as well as the posterior nerve-roots, were normal. The anterior nerve-roots, on the contrary, were degenerated for the first three pairs below the section; and the same alteration continued, diminishing gradually in amount, for the fourth, fifth, and sixth pairs. Thus there may be, within the columns of the cord, as well as in its nerve-roots, centripetal or centrifugal degenerations, affecting exclusively particular nervous tracts.

This discovery corresponded with certain pathological observations in man, made about the same time by Türck, and communicated by him to the Academy of Sciences at Vienna.[2] They were cases in which post-mortem examination revealed an old morbid deposit in the brain, together with structural alteration of the spinal cord on the opposite side. This alteration consisted mainly in a production of

---

[1] Comptes rendus de l'Académie des Sciences Paris, 1852. Tome xxxv. p. 301.
[2] Sitzungs-berichte der kaiserlichen Akademie der Wissenschaften. Wien, 1851, p. 288.

abnormal granular cells, occupying the tissues of the cord, and resembling those met with about the morbid deposit in the brain. The most striking feature of the alterations in question was that they occupied definite longitudinal tracts, the rest of the cord being normal. By examining successive sections at different levels, it became evident, that, from the lesion in the brain to the lower end of the spinal cord, only certain bundles of fibres were affected; and these bundles corresponded in situation with those which could be demonstrated by dissection in normal specimens of the cord, after hardening in alcohol.

Observing this correspondence, in certain cases, between the degenerated tracts and parts already known, the author was led to apply the same means of observation to the study of anatomical relations not so well understood. The attempt proved successful. The first fact which it brought out was a distinction between the anterior and posterior segments of the lateral column. No such distinction can be shown by ordinary dissection; most of the nerve-fibres in both segments running in a direction parallel with each other, and between origins and terminations too distant to be traced. But, in some of the cases mentioned, the morbid alteration of the lateral column was found only in its posterior part, the anterior part remaining sound; and this difference was so constant, in sections made at various levels, as to leave no doubt that there must be some anatomical distinction between the two segments of the column, extending through the greater part of its length.

It was also seen, in these cases, that, on approaching the upper end of the cord, the degenerated tract shifted its position forward and inward, and, at the medulla oblongata, crossed into the anterior pyramid of the opposite side. Thence it extended upward through the pons Varolii, where the distinction was extremely evident, the transverse bundles of the pons being free from alteration, while its longitudinal bundles were deeply affected. It appeared, therefore, that the degenerated tracts had been invaded by a special alteration confined to their own limits, and extending continuously from the seat of the brain-lesion, through the crus cerebri, the pons, and the anterior pyramid of the same side, and thence, through the posterior part of the lateral column on the opposite side, to the lower end of the spinal cord.

This showed that there was a longitudinal tract of nerve-fibres, reaching for long distances in the cerebro-spinal axis, and having, so to speak, its own centre of nutrition; and that, when this centre of nutrition was disturbed or disorganized by local lesion, the subsequent degeneration might extend throughout the tract in question, while others in its immediate contiguity remained sound. The cases observed by Türck were accompanied by more or less hemiplegia; and while the brain-lesion, as well as the degeneration of the pons and anterior pyramid, were, as usual, on the opposite side, that in the spinal cord was on the same side with the paralysis. Every thing showed the degenerated portions in the spinal cord to be continuous with the opposite anterior pyramids and their extension in the crura cerebri, and consequently to represent the motor tracts of the cerebro-spinal axis. But while Türck followed the crossing of these tracts at the decussation of the pyramids, he also showed that in certain cases their crossing was incomplete; that is, although the bulk of a degenerated anterior pyramid crosses at the level of the decussation, and re-appears in the lateral column of the opposite side, a remnant of its fibres continue their downward course for some distance on the same side. These direct columns, visible more or less constantly in the cervical portion of the cord, are still known by the name of the "columns of Türck."

But his most important results were those indicating the difference, in anatomical connection, between the anterior and posterior parts of the lateral column. These two parts were distinguishable from each other by the granular deposit with which one was infiltrated, while the other remained normal; and the deposit appeared with such regularity at the same spot in successive sections, and crossed so distinctly at the decussation of the pyramids, as to show that it followed the course of an associated bundle of nerve-fibres. As this bundle represented, in the cord, the continuation of the anterior pyramid above, the whole tract, from end to end, afterward received the name of the "pyramidal tract." The first indication of the exact locality of this tract in the lateral columns was derived from the sections of Türck.

The same author made an incidental observation, afterward seen to be very important, in regard to degenerations from local lesion of the cord itself. They were instances of paraplegia from compression or disorganization of the spinal cord at one point. In these cases the portion of the cord above the lesion, while mainly free from degeneration, nevertheless contained single degenerated tracts, always identical in situation, extending upward to the medulla and pons. But these tracts were not the same with those found degenerated after a brain-lesion; and the latter were entirely sound in the above-mentioned cases of injury to the cord. If one portion of the spinal cord, therefore, can degenerate in consequence of a lesion situated above it, another, and an entirely different portion, may be affected by degenerations coming from below.

So far, the knowledge of degenerations within the spinal

cord had been mainly acquired from pathological observations in man. But, about ten years ago, Goltz, in Strasburg, undertook a series of investigations on the functions of the lumbar portion of the spinal cord, with regard to its influence on the sexual and urinary organs. His experiments were performed on young dogs, by dividing the cord transversely at the lower end of the dorsal region, and then preserving the animals long enough to study the permanent effects of the operation. By gradual improvements in the selection of animals, the operative procedure, and the subsequent treatment, he became very successful. The cord was divided without serious injury to the vertebral column or the external parts. Dogs of the proper age and organization were not dangerously affected by the operation, sometimes taking food within the first twenty-four hours, and generally regaining their appetite in a few days. The wounds healed in two or three weeks; and some of the animals were preserved several months, remaining paraplegic, but otherwise in good condition.

This offered a valuable material for the study of degenerations in the spinal cord, and it was utilized for that purpose by Schieferdecker,[1] the assistant of Goltz, who was familiar with the history and condition of the animals employed.

On examining microscopic sections of the cord, in these cases, the granular degeneration of the pyramidal tracts was recognized. It was present in every instance, and had invariably taken a descending course from the point of section to the lower extremity of the cord. This portion of the spinal cord, therefore, had become altered in the same way as a divided spinal nerve, that is, from the centre toward the periphery; and the degeneration of the pyram-

[1] Archiv für pathologische Anatomie und Physiologie. Berlin, 1876. Band lxvii. p. 542.

idal tract, whether it be due to a cerebral lesion, or to compression or section of the cord itself, follows the same direction.

On the other hand, when a similar degeneration takes place in the posterior columns of the cord, it is always ascending. We have now the experience of Charcot[1] to corroborate the earlier assertion of Türck, that such posterior degenerations invariably extend from the point of injury or disease upward toward the brain, and never in a downward direction. By this means the ascending and descending degenerations of the spinal cord were established with the same certainty as the centripetal and centrifugal degenerations of the nerve-roots, and, like them, were shown to occupy separate tracts, easily distinguishable from each other.

These facts greatly increased the confidence of physiologists in the value of degenerations, as indicating the continuity and destination of nerve-fibres. It was evident from the first that a divided nerve-fibre degenerates because of its severance from some special nerve-centre, and, once cut off from this centre, it becomes affected throughout its length. When we see, in the spinal nerves, complete degenerations strictly limited to the ramifications of a single trunk, and similar alterations in the spinal cord equally confined to particular tracts, we can hardly refuse the conviction that these tracts also consist of continuous nerve-fibres following a common direction. Waller's experiments on the nerve-roots show, that, when such a group of fibres meets a ganglion or centre of gray substance interposed in its course, the degeneration does not pass this point. Either the injured fibres receive an

[1] Leçons sur les Localisations dans les maladies du Cerveau et de la Moelle épinière. Paris, 1880, p. 247.

accession of nutritive energy from the ganglion, or else they terminate in its substance and are replaced by new fibres originating from its gray matter. This gives a special significance to the fact observed by Charcot[1] in regard to disease of the pyramidal tracts. These tracts are manifestly the channels for voluntary motion in the spinal cord. The paralysis caused by either their morbid alteration or their experimental division shows that they are the paths followed by voluntary motor impulses from the brain to the anterior nerve-roots. But descending degenerations of the pyramidal tract, however complete in the spinal cord, do not usually extend to the motor nerves or nerve-roots. This is easily explained on the assumption that the fibres of the pyramidal tract terminate in the gray substance of the anterior horn, while those of the anterior root have their origin at the same point. The nerve-root therefore degenerates only when divided beyond its emergence from the anterior horn.

Further study of the posterior columns led to the discovery of two different kinds of degeneration, as well as to a distinction between different parts of these columns. Hitherto the morbid alteration of nervous tracts had appeared only as the consequence of a local lesion in either the nerves, spinal cord, or brain; and it extended upward or downward from its point of origin, as if propagated in certain directions along the natural route of the nerve-fibres. These alterations of structure were therefore known as "secondary degenerations," because resulting from some primary affection in a different locality. But it appeared that certain tracts in the spinal cord might also be the seat of primary degenerations, originating within their own substance independently of lesions elsewhere. As these alter-

[1] Leçons sur les maladies du Système Nerveux. Paris, 1877. Tome ii. p. 219.

ations often followed a similar course to those of secondary origin, occupying only their own tracts or systems of fibres, they received the name of "systematic degenerations."

The posterior columns of the cord were liable to both secondary and systematic degenerations. But these two affections occupied different regions, thus showing an anatomical distinction, in the posterior column, like that already established between the front and back parts of the lateral column.

In the posterior column of the cord, the inner part, next the median line, is a narrow band, visibly marked off from the remainder in the cervical region by a slight furrow on the surface. This is the *funiculis gracilis*, or the "column of Goll." At the medulla oblongata it diverges from the median line obliquely upward, forming, on the inner border of the restiform bodies, the so-called "posterior pyramids." The columns of Goll are the seat of the ascending degenerations noticed by Türck. When they are affected in this way, their alteration extends unbroken from its starting-point often quite to the level of the medulla oblongata, whence it is inferred that they are composed throughout of continuous fibres.

But the remainder of the posterior column is affected by degenerations which have a different form and different results. This portion is included between the column of Goll and the posterior nerve-roots, and forms the outer part of the posterior column. It has not received a distinct name, but is important, owing to its connection with locomotor ataxia. It is regarded as made up of comparatively short fibres, originating and terminating in succession along the cord, for the reason that its secondary degenerations are never propagated to any considerable distance, extending at most only two or three centimetres above their origin.' But it is liable to systematic alterations, mainly sclerosis, invading a large portion of its extent ; and, when such alterations occur, they are accompanied by symptoms of locomotor ataxia, never produced by structural disease in other parts. According to the observations of Charcot,[1] the columns of Goll may be completely degenerated without causing any signs of ataxia, while sclerosis of the outer part of the posterior columns is always accompanied by ataxic symptoms; and these symptoms are in proportion to the extent of the degeneration.

Lastly, the anatomical study of nervous tracts, by the aid of secondary degenerations, has been carried into the brain itself. There is already a reasonable certainty of the extension of the pyramidal tracts through the anterior pyramids, the pons Varolii, and the crura cerebri; and the fibres of the crus cerebri are manifestly succeeded by the diverging expansions of the internal capsule and the corona radiata. These expansions are, beyond question, for the most part, means of communication between the cerebral convolutions and the ganglia at the base of the brain, while the fibres of the crus cerebri connect the ganglia with the spinal cord below. But, beside these interrupted communications, are there also *direct* fibres, running continuously from the motor centres in the cerebral cortex, through the corona radiata and internal capsule, into the pyramidal tract? The centres of motion about the fissure of Rolando might produce muscular contractions, when galvanized, either through the intervention of the cerebral ganglia, or, more directly, by means of continuous fibres. Such direct

---

[1] Leçons sur les maladies du Système Nerveux. Paris, 1877. Tome ii. p. 11. Leçons sur les Localisations dans les maladies du Cerveau et de la Moelle épinière. Paris, 1880, p. 259.

tracts have been sometimes imperfectly recognized in the examination of hardened specimens; but they have never been clearly demonstrated by this method, owing to the complicated interlacement of fibres in the upper part of the internal capsule. Their existence is mainly inferred from the course of descending degenerations in this region. According to Charcot, destructive lesions of the cortex, in the anterior and posterior central convolutions, give rise to degenerations in the white substance, which pass downward through the internal capsule, crura cerebri, anterior pyramids, and lateral columns of the cord. These degenerations may exist without lesion of the cerebral ganglia; and they are not produced by morbid alterations in other parts of the brain than those about the fissure of Rolando. Similar observations, during a period of fifteen years, point with much significance to a continuity of fibrous structure throughout this tract. If we remember the uniform progress of discovery thus far made by these investigations,— first in the nerves, then in the nerve-roots, then in the columns of the spinal cord and medulla oblongata,—it does not seem presumptuous to expect, from their further prosecution, a corresponding success in the more difficult exploration of the brain.

The study of nervous degenerations has already borne fruit of great value. Like many other researches of similar character, it was at first a matter of purely scientific interest, without any apparent bearing on the treatment or even the pathology of disease. But it has grown in importance with every additional observation, and it is now an indispensable aid in investigating the morbid affections of the nervous system. It has extended our anatomical knowledge far beyond its previous limits. It has demonstrated the existence of nervous connections which could hardly have been detected in any other way. It has shown the relation between certain forms of paralysis and the altered structure of particular nervous tracts, and it seems likely to yield results of still wider application before its capacity is exhausted. It began with the degenerating fibres of a divided sciatic nerve, and it has now reached the intricate region of the white substance of the cerebral hemispheres.

Sir Charles Bell's THEORY OF THE NERVOUS SYSTEM is of so recent a date, comparatively speaking, that its traces have not entirely disappeared from our nomenclature. The student will still find, among the synonymes in some anatomical text-books, such names as the "Respiratory nerve of the face," or the "Superior and External respiratory nerves of Bell." But the system of which these names formed a part has become so obsolete, that their significance is hardly remembered; and even the distinctive features of the theory itself have been forgotten in the discussion of more important topics connected with the nervous system.

Among the eminent medical men of the present century, perhaps none have attracted more cordial admiration than Sir Charles Bell. His earnest and strictly professional ambition, his restless mental activity, his versatility of talent, and his genial and enthusiastic disposition, secured him a rapid success and a position of acknowledged superiority. "Anatomist, surgeon, author, artist, and critic," he taught anatomy by lectures and demonstrations, was the equal of Cooper and Cline in practical surgery, published his magnificent works on the "Operations of Surgery" and the "Anatomy of Expression," illustrated by himself, and created the reputation of the Middlesex Hospital as a centre of medical instruction. When he received the order of knighthood, in 1831, at the same time with Herschel, Babbage, and Brewster, he thought more of his companions than of his title. "The batch," he said, "makes it respectable." He always retained a strong attachment to his early friends and family relatives, and to the pleasures and sports of a country life; and, if he experienced some disappointment as to the success of his scientific opinions, this was fully compensated by his general popularity and his high professional reputation.

Bell's attention was especially attracted to the nervous system in 1807, a few years after his arrival in London, and while occupied as lecturer in a private medical school established by himself. In his demonstrations he was struck with the apparent complexity of the nervous apparatus, and the difficulty of presenting it fairly to the comprehension of his audience. His mind dwelt persistently on this subject, with a view of reaching some more intelligible arrangement of what seemed so confused and disconnected. His thoughts were first turned to the anatomy of the brain, which, he says, at that time "occupied his head almost entirely."[1] He puzzled over the varied form and structure of the cerebrum and cerebellum, the processes, protuberances, and ganglia at the base of the brain, and the connections and course of the cranial nerves. He tried to reduce these complicated parts to something like order in his mind, and to divine the secret of their anatomical relations.

This was the origin of his new system in cerebral anatomy; and, when once it had assumed a distinct form, it grew in importance, and took entire possession of his con-

[1] Letters of Sir Charles Bell. Selected from his correspondence with his brother, George Joseph Bell. London, 1870, p. 117.

viction and imagination. He made it the subject of his lectures, and speaks of it in his correspondence with the greatest enthusiasm. "I hinted to you," he says in a letter to his brother in Edinburgh, "that I was *burning*, or on the eve of a grand discovery. . . . I really think this new anatomy of the brain will strike more than the discovery of the lymphatics being absorbents. . . . My object is to lecture it, to make the town ring with it, as it is the only new thing that has appeared in anatomy since the days of Hunter, and, if I make it out, as interesting as the circulation or the doctrine of absorption." [1]

Up to this time there is no trace of any experimental investigations by Bell in regard to the subject. His system was wholly inferential, based on the facts of anatomical structure. But a year or two later he felt the necessity of making some experiments, especially as he had formed the intention of printing his views in a pamphlet for private distribution. He hoped by this means to elicit the opinion of the representative men of the profession, and to see whether his innovations would be likely to meet with approval.

Accordingly, in 1810,[2] he records having made two experiments with a satisfactory result; so much so, that he believes he is now about to establish his system on facts, — "the most important that have been discovered in the history of the science." And in 1811 his essay was printed, under the title "Idea of a New Anatomy of the Brain; submitted for the Observations of his Friends, by Charles Bell, F.R.S.E." [3]

The idea developed in this essay, and the importance

---

[1] Letters of Sir Charles Bell, pp. 117, 118. [2] Ibid., p. 170.
[3] Reprinted in the Journal of Anatomy and Physiology. Cambridge and London, 1869. Vol. iii. p. 153.

attached to it in the mind of its author, can be best understood from Bell's own account of its origin and growth. In a letter to his brother a year before, he details his plan of investigation, and the object to be accomplished. He had already reached the conviction, on anatomical grounds, that the several divisions of the encephalon were different from each other in function, and especially that a wide distinction of this kind existed between the cerebrum and the cerebellum. But how gain access to these deep-seated and vascular parts, so as to experiment upon them? This seemed too difficult an undertaking. It might be possible, however, to expose, for that purpose, the spinal cord and its nerve-roots; and this is what he accordingly did. "It occurred to me," he says, "that, as there were four grand divisions of the brain, so were there four grand divisions of the spinal marrow: first, a lateral division, then a division into the back and fore part. Next it occurred to me that all the spinal nerves had, within the sheath of the spinal marrow, two roots, — one from the back part, another from before. Whenever this occurred to me, I thought that I had obtained *a method of inquiry into the functions of the parts of the brain.*"[1]

That is to say, experiments on the anterior columns of the spinal cord, or the anterior nerve-roots, would throw light on the functions of the cerebrum, with which they were anatomically connected. Experiments on the posterior columns of the cord, or the posterior nerve-roots, would indicate the function of the cerebellum, to which, he thought, these parts belonged.

These are the experiments described in the "Idea of a New Anatomy of the Brain," and this is the meaning attributed to them by its author. He found that irritation

[1] Letters of Sir Charles Bell, p. 170.

applied to the anterior part of the spinal cord produced contraction in the voluntary muscles, while a similar injury inflicted on its posterior part did not do so. On laying bare the roots of the spinal nerves, he could cut across the posterior fasciculus without convulsing the muscles; but touching the anterior fasciculus at once produced convulsions. "Such were my reasons," he says, "for concluding that the cerebrum and cerebellum were parts distinct in function, and that every nerve possessing a double function obtained that by having a double root."[1]

The distinction made by Bell between the two divisions of the encephalon is this: the cerebrum is the organ of the mind, of conscious sensation and volition: consequently, the anterior roots of the spinal nerves, being connected with the cerebrum through the anterior part of the spinal cord, are subservient to sensation and volition, and, when artificially irritated, cause contraction in the voluntary muscles. The cerebellum, on the other hand, presides over the secret or unconscious operations of the bodily frame, as in the vital action of the internal organs: consequently, irritation of the posterior columns of the cord, or the posterior nerve-roots, has no perceptible effect on the voluntary muscles. A spinal nerve accordingly possesses two roots, because it performs two sets of functions,—one, those of consciousness and volition, derived from the cerebrum; the other, those of unconscious vital influences, derived from the cerebellum. The cranial nerves, however, have for the most part only a single root, and exercise but one set of functions, corresponding to their place of origin. Thus "the eighth nerve (par vagum) is from the (back) portion of the medulla oblongata which belongs to the cerebellum; the ninth nerve (sublingual) comes from the (front) portion which belongs

[1] Reprint in the Journal of Anatomy and Physiology, vol. iii. p. 161.

to the cerebrum. The first is a nerve of the class called *vital* nerves, controlling secretly the operations of the body; the last is the motor nerve of the tongue, and is an instrument of volition."[1] This is the substance of the views contained in the essay of 1811.

Bell was much disappointed in the impression produced by this pamphlet. It turned out an "unpropitious experiment;" for it "excited no criticism, and threatened to stifle the enthusiasm of the author." Bell himself says regretfully that this announcement, from which he expected so much, "failed to draw one encouraging sentence from medical men." But we can hardly blame them very much for this apathy. The systematic ideas of the author, however attractive to himself, might not have the same interest for others; and his experimental proofs were certainly insufficient to command belief on their own merits. Indeed, we now know that they do not really bear the significance which he attached to them. The anterior and posterior nerve-roots are not connected respectively with the cerebrum and cerebellum, and we cannot use them in our experiments to determine the separate functions of the brain. It was, no doubt, a suspicion of this physiological *non sequitur* that prevented the pamphlet from being appreciated as its author had hoped.

But, notwithstanding this discouragement, Bell afterward returned to the subject with new vigor, and soon came to regard his former work as only the introduction to a more extended system. In his first essay, the brain was the main subject of research, the nerves and their distributions being only secondary; but, in his theory as afterward developed, the nerves and their distributions occupied the prominent place. This matter, in 1814, was already, he says, "ripen-

[1] Reprint in the Journal of Anatomy and Physiology, vol. iii. p. 162.

ing in his head;" and in his lectures on the nerves for that year he promises to "lay open a fine system." Like his anatomy of the brain, the system increased in importance the more he thought about it. Year by year he enlarged its proportions, and seemed to anticipate from it greater results than ever. He gradually brought it into the form of a series of communications to the Royal Society. In 1819 he regarded it as nearly complete, and fully worth the labor it had cost, containing, as he believed, the materials of a grand system, destined to "revolutionize all we know of this part of anatomy, more than the discovery of the circulation of the blood." By the time the first paper was ready for presentation he was abundantly satisfied with it, and declared that he had made "a greater discovery than ever was made by any one man in anatomy."

In this communication,[1] Bell presented the main principles of his new arrangement of the nervous system, with special reference to the nerves of the face. The paper had been prepared by its author with much care, and he offered it to the Royal Society as the result of several years of thought and investigation.

This time he had no reason to be disappointed, as the essay attracted everywhere attention and approval. In speaking of a reception which he attended soon afterward, "My paper," he says, "has done me as much good as if I had bought a new blue coat and figured French black silk waistcoat. One gentleman called it the first discovery of the age." He received a number of complimentary letters in regard to it, which, he says, were enough to show that he was "not a visionary" on the subject. "It will hereafter," he adds, "put me beside Harvey."[2]

[1] Philosophical Transactions of the Royal Society. London, 1821, p. 398.
[2] Letters of Sir Charles Bell, p. 272.

This paper of 1821 well deserved the praise bestowed upon it, for it contained the announcement of Bell's most lasting and undisputed discovery; namely, that the facial nerve, or *portio dura* of the seventh pair, was not a nerve of sensibility, but a channel for muscular action and the nervous medium of expression in the face. The importance of this discovery, both scientific and practical, was abundantly manifest; and it was sustained by direct and satisfactory experiments on animals, as well as by observations on man. It is evident that Bell did not get at the whole truth in regard to this nerve; for he represented its action as limited to the involuntary movements of respiration and expression, and considered the voluntary motions of the face and lips as still provided for by the fifth pair. But it is very seldom that all the details of so complex a subject are mastered by one person, or at one time; and the primary facts of the motor character of the facial nerve and its insensibility, as contrasted with the extreme sensibility of the fifth pair, were then fully established on incontestable evidence.

Bell, however, was much more interested in the general plan of his classification than in any of its details, which he valued mainly as illustrations or proofs of the system as a whole. He continued its development in his communications to the Royal Society; and it was afterward embodied, with some necessary modifications, in his "Nervous System of the Human Body," published in 1830.

Bell's conception of the arrangement of the nerves was this: All the nerves of the cerebro-spinal system were divided into two great classes or groups. The first group was that of the *regular, original,* or *symmetrical* nerves; namely, the thirty-one pairs of spinal nerves and the fifth pair of cranial nerves. All these nerves arise by two roots, on one of which is a ganglion; they pass out regularly to successive divisions of the body; they are all subservient to voluntary motion and common sensibility; they are distributed to every part of the frame, but are symmetrical and simple in their arrangement. The second group was made up of the *irregular* or *superadded* nerves. These do not arise by double roots, and have no ganglia at their origins: they come off from the sides of the medulla oblongata and upper part of the spinal cord; and, instead of being distributed in regular order to successive parts of the frame, are sent to remote organs, which they combine in functional activity. They therefore have the character of being superadded, for a special purpose, to the original system of regular nerves. They constitute the great group of *respiratory nerves*, which are distributed to a variety of organs, and, instead of going out on each side straight to their destination, often pass through the body in a longitudinal or oblique direction. They are, —

First, The facial, portio dura of the seventh pair, respiratory nerve of the face.

Second, The glosso-pharyngeal nerve, distributed to the tongue and pharynx.

Third, The par vagum, or pneumogastric, the nerve of the lungs, heart, and stomach.

Fourth, The spinal accessory, or superior respiratory nerve of the trunk, distributed to the sterno-mastoid and trapezius muscles.

Fifth, The sublingual, or the nerve of articulation.

Sixth, The phrenic, or great internal respiratory nerve.

Seventh, The long thoracic, or external respiratory nerve of the trunk.

Eighth, The fourth nerve, or patheticus, distributed to the superior oblique muscle of the eyeball.

These are the nerves which produce the appearance of irregularity or confusion in the nervous system; because they often cross the track of the symmetrical nerves, and are distributed to organs already supplied by them.

This last fact gives the key to the functional peculiarity of the superadded nerves. According to Bell, an organ which performs only a single function has but one nerve supplied to it; and, whenever an organ receives nerves from two or more sources, it is because it performs as many different functions. This explains why certain muscles, like the sterno-mastoid and trapezius, already supplied with nerves from the regular or symmetrical system, also receive branches from the respiratory system. The former nerves enable them to execute the ordinary acts of voluntary motion: the latter bring them into occasional involuntary unison with the act of respiration. This is illustrated in a still more striking manner by the two nerves distributed to the face; namely, the fifth pair, supplying the face with sensation and voluntary motion,[1] and the facial nerve, or seventh, a superadded nerve, which controls the involuntary motions of the face in breathing,[2] or, as he says, "when the muscles of these parts are in associated action with the other organs of respiration."

The respiratory nerves, however, come into play, not only in the motions of respiration proper, but in all those connected with the entrance and exit of air from the lungs; as in speaking, singing, coughing, gasping, or sneezing. They are also the agents in all movements of expression, or the emotional affections of the frame. This is especially marked in the seventh pair, which regulates the involuntary changes of expression in the face. But it applies also to

[1] Philosophical Transactions, 1821, pp. 410, 411, 413, 417; and 1822, p. 284.
[2] Ibid., 1821, pp. 406, 410, 414.

the manifestations of emotion in other parts; as in the hurried respiration of anxiety, the retardation or quickening of the heart, and the sobbing of the diaphragm in grief, excitement, or depression. Under every strong impulse, the central organs of respiration are stimulated or disturbed, and thus bring into associated action all their subordinate parts as organs of expression.

Furthermore, the nerves of respiration, wherever they may be, have greater vitality than other nerves; for the movements connected with this function continue to be performed after sensibility and volition have disappeared. Thus these nerves form a great system by themselves, distinguished from the ordinary or regular nerves by their mode of origin, their functions, their distribution, and their vitality.

Evidently there was much of real value in these observations. They placed in strong relief the fact of associated respiratory motions. The expansion and collapse of the nostrils or lips, accompanying the action of the thorax, are occasional in man, constant in some animals, and are always in proportion to the intensity of respiration. They are truly respiratory motions, as much so as those of the diaphragm. The sterno-mastoid and trapezius muscles are also brought into action whenever breathing is impeded or laborious. The classification which Bell constructed on this basis, and illustrated with so much ingenuity, has a certain charm about it, and is plainly marked with the talent and originality of its author. But, even as a system, its joints do not all bear inspection; and it contained from the outset certain defects, which became after a time plainly perceptible.

Bell showed by his experiments and observations that the seventh nerve controls the involuntary motions of expression and respiration in the face, and that these motions are abolished by its division or injury. If he had accepted this fact at its own value, and had assumed nothing further, he would have been secure; but he believed, in accordance with his system, that these involuntary movements were the only ones under control of the seventh nerve, and that sensibility and volition were provided for by the fifth pair. This idea influenced his mind to such an extent, that he was led to conclude, after division of the fifth pair alone, that the voluntary movements of the lips were paralyzed, while those of respiration continued.[1] Magendie, on the other hand, who believed in nothing but direct experiment, repeated this operation on the fifth pair, and declared that he could not perceive any paralysis of the lips produced in consequence.[2] We now know that he was right, and that the fifth pair does not animate any of the superficial muscles of the face. That being true, according to Bell's principle, there ought to be *two* other nerves distributed to this part, — one for its voluntary, the other for its respiratory functions; but, in fact, it has only the seventh pair, which serves for both.

Bell's classification of the respiratory nerves, with regard to their origin, is by no means perfect in its details. These nerves were said to differ from the others in having no ganglia at their roots, and in originating from an intermediate tract on the sides of the medulla and spinal cord. But both the pneumogastric and glosso-pharyngeal nerves have ganglia at their roots, which are as real, though not quite so conspicuous, as the Gasserian ganglion of the fifth pair. Again: the origin of nerve-roots, in a nearly continuous line, from the lateral tract of the medulla and spinal

[1] Philosophical Transactions, 1821, p. 413.
[2] Journal de Physiologie. Année 1821. Tome i. p. 387.

cord, is certainly a marked feature in the glosso-pharyngeal, pneumogastric, and spinal accessory nerves. On the other hand, it cannot be asserted, with any plausibility, of the facial, sublingual, and patheticus.

The last-named nerve was associated with the respiratory system, because it was regarded as causing the upward rolling of the eyeball in involuntary movements and emotional expressions. Every thing connected with this subject had for Bell a great attraction. His book on the "Anatomy of Expression" was his earliest publication in London, and first called attention to him as a man of ability. He gave lectures on the same topic to professional artists; and the exercise of his talent in painting and modelling was always for him a source of pleasure in connection with his anatomical pursuits. This probably accounts for the rather forced amalgamation, in his theory, of the functions of respiration and expression.

But how does it happen that Bell omitted from his respiratory system the *intercostal* muscles and their nerves? One would think they had some claim to be considered as part of this great apparatus. They are constantly at work, and keep time with the diaphragm and lungs more steadily than any other outlying organs of respiration. But the intercostals belong to the regular symmetrical system of nerves, originating from the spinal cord with double roots and all the other features of their class. The author ignores this difficulty almost entirely, or alludes to it only so far as to surmise that the branches of these nerves which influence respiration are, "in all probability," derived from the same lateral tract of the spinal cord.[1] But there is not the least evidence adduced in support of this probability, and the assumption looks very much like reasoning in a circle.

[1] Philosophical Transactions, 1822, p. 287.

The controversy once carried on as to the exact significance of Sir Charles Bell's doctrine and discoveries was due, in great measure, to the singular obscurity of his style, and his indefinite manner of statement. These defects were habitual with him as a writer, and in many instances are extremely marked. There are few pieces of harder reading than the preface and introduction to his "Nervous System of the Human Body."[1] In attempting to account for this obscurity, the reader is sometimes in doubt whether Bell really has any distinct idea to communicate, or whether he voluntarily stops short of its complete expression. One of these puzzling passages is the following. After speaking of the par vagum and the spinal accessory nerve, he says (p. 47), "Directed in the next place to the portio dura, I wished to answer the question, 'Why does the nerve which supplies certain muscles of the face take an origin and a course different from the fifth nerve, destined to the same parts?' *Guided by these considerations in my experiments, by inference I concluded*, that, on cutting across this nerve, all the motions of the face connected with respiration ceased, and that it had the origin we see, and took its course with the respiratory nerves, because it was necessary for the association of the muscles of the nostrils, cheek, and lips, with the other muscles used in breathing, speaking, etc." It is certainly impossible to determine, from the phraseology of this sentence, whether it contains the statement of an experimental fact, or is only the expression of a hypothetical surmise.

Having enumerated again the various organs of respiration, he says (p. 49), "It appears, then, that it is the distance and irregular position of the eye, nostril, mouth, throat, and larynx, and muscles of the neck, which require

[1] First edition. London, 1830.

these diverging and apparently irregular nerves, to connect them with the act of respiration, and without which they would have possessed no more attributes than the nerves of the limbs; that is to say, sensibility and motion."

There is nothing in the context to indicate from what point the eye, nostrils, mouth, throat, and larynx are "distant," or why they should be considered as more "irregular" in position than the limbs.

One of the most curious instances of this peculiarity is a passage in one of Bell's earlier letters, describing his new system of the anatomy of the brain; and I think it hardly probable that any one now present will be able to understand exactly what he means by it. "I consider," he says,[1] "the organs of the outward senses as forming a distinct class of nerves from the other. I trace them to corresponding parts of the brain, totally distinct from the origins of the others. I take five tubercles within the brain as the internal senses. I trace the nerves of the nose, eye, ear, and tongue to these. Here I see established connections. Then the great mass of the brain receives processes from these central tubercles. Again · the greater mass of the cerebrum sends down processes, or crura, which give off all the common nerves of voluntary motion, etc. I establish thus a kind of circulation, as it were. In this inquiry I describe many new connections. The whole opens up in a new and simple light: the nerves take a simple arrangement; the parts have appropriate nerves; and the whole accords with the phenomena of the pathology, *and is supported by interesting views.*"

This indefinite style of expression extends, as in one of the passages just quoted, even to the description of experiments and their results. It is usually thought essential in

[1] Letters of Sir Charles Bell, p. 117.

reporting experiments, especially if bearing on new questions in physiology, to state all the particulars, so that the reader may form for himself some estimate of their value. Bell habitually neglects this useful precaution. He generally omits all mention of the details of the operative procedure, often that of the species of animal employed, and sometimes even whether it were living or dead at the time of the experiment. The consequence is, that doubts have been entertained on these points in regard to some of his most important investigations.

The truth is, Bell had no real faith in experimentation as a source of knowledge. He preferred to make his deductions, in the first place, from the details of anatomical structure; employing experiments afterward to "prove" or "confirm" them, or to impress his convictions on the minds of others. This is abundantly evident from his own express declaration in various parts of his works, as well as from the method pursued in originating and carrying out his investigations. With him, the conception of a system was the beginning and the end of the mental process: its experimental evidence was only an intermediate episode. For this reason, Bell would hardly be considered, at the present day, as a physiologist. Indeed, he never speaks of himself as such, but always as an "anatomist," whose whole pleasure is in "investigating structure;" and even his new doctrines and ideas he designates as "discoveries in anatomy."

This explains, in some degree, his undue estimate of the two methods of inquiry. He had so high a regard for the results of anatomical research, that he believed them capable of also solving questions in physiology. Already, in 1808, three years before printing his "Idea of a New Anatomy of the Brain," he "is sure," he says, "that he is correct;" and that was before he had made a single experiment. And long afterward he declares that the "few experiments" which he has made "were directed only to the verification of the fundamental principles on which the system is established." But the result of this method has been exactly the reverse of what he anticipated. The experimental observations, which he undervalued, form the only part of his system which has remained; while the attractive and harmonious classification, which he constructed with so much ingenuity on the basis of anatomical deduction, has now lost its importance and almost its place in medical literature.

The value of the experimental method in medical science cannot be measured by the direct results of any particular discovery. It is true that these discoveries are often of great importance, and increase largely the fund of our medical knowledge. They possess, moreover, a vitality which distinguishes them in a marked degree from the ephemeral products of scientific hypothesis. When a hypothetical system has served its time, it disappears, and is replaced by a different one; but the knowledge derived from experiment remains serviceable long after its novelty has passed away. The experimental observations of Galen on the recurrent laryngeal nerves and on the functions of the arteries as blood-vessels are as conclusive now as when he first made them, and retain at this day their full value. They have lasted for over seventeen centuries, and have survived, during that time, all the fluctuating medical systems of solidism and fluidism, of animism, vitalism, archeism, and iatro-mechanism. The information which they imparted was a reality, and is neither destroyed nor impaired by the lapse of time.

But the benefits of an experimental discovery extend far beyond its immediate limits. When a new fact has been established in anatomy or physiology, it is impossible to say what bearing it will have on other facts not yet discovered; and in the history of medicine there are many instances of curious observations, which remained imperfect and apparently barren until subsequent discoveries invested them with an unexpected importance. It is evident that Galen's detection of the arterial blood-current in the second century was a necessary preliminary to Harvey's complete discovery of the circulation in the seventeenth; and the lymphatic vessels, discovered in 1651, were thought to be the agents of a slow and insignificant absorption, until Colin found, two hundred years later, that the daily quantity of fluids passing through the thoracic duct was from four to ten per cent of the entire bodily weight. It is almost certain that a genuine investigation, however isolated, will produce, some time or other, its legitimate fruit. But, to secure this object, there is one essential requisite; and that is, that experimental research be followed and cultivated for its own results, whatever they may be, without demanding immediate returns of a kind that can be designated beforehand, and with entire confidence in the substantial value of this method, which has always been, and will always be, the only source of permanent improvement in medical science.

www.ingramcontent.com/pod-product-compliance
Lightning Source LLC
Chambersburg PA
CBHW020143170426
43199CB00010B/862